Principles of Quality Costs

Fifth Edition

Douglas C. Wood and W. Frazier Pruitt

ASQ Quality Press
Milwaukee, Wisconsin

Principles of Quality Costs, Fifth Edition
Doug Wood and Frazier Pruitt
American Society for Quality, Quality Press, Milwaukee 53203

Publisher's Cataloging-in-Publication Data

Names: Wood, Douglas C., 1955–, author. | Pruitt, W. Frazier, 1986–, author.
Title: Principles of quality costs , fifth edition / by Doug Wood and Frazier Pruitt.
Description: Includes bibliographical references. | Milwaukee, WI: Quality Press, 2026.
Identifiers: LCCN: 2026935921 | ISBN: 9781636942117 (paperback) | 9781636942131 (pdf) | 9781636942148 (epub)
Subjects: LCSH Quality control—Costs—Case studies. | Service industries—Quality control—Costs—Case studies. | Manufactures—Quality control—Costs—Case studies. | BISAC BUSINESS & ECONOMICS / Quality Control | BUSINESS & ECONOMICS / Total Quality Management | BUSINESS & ECONOMICS / Leadership | BUSINESS & ECONOMICS / Organizational Behavior
Classification: LCC TS156 .W66 2026 | DDC 658.562—dc23

ASQ advances individual, organizational, and community excellence worldwide through learning, quality improvement, and knowledge exchange.

Bookstores, wholesalers, schools, libraries, businesses, and organizations: Quality Press books are available at quantity discounts for bulk purchases for business, trade, or educational uses. For more information, please contact Quality Press at 800-248-1946 or books@asq.org.

To place orders or browse the selection of all Quality Press titles, visit our website at: http://www.asq.org/quality-press.

30 29 28 27 26 LS 7 6 5 4 3 2 1

∞ Printed on acid-free paper

Quality Press
600 N. Plankinton Ave.
Milwaukee, WI 53203-2914
Email: books@asq.org
ASQ Excellence Through Quality™

Dedication

To my wife for her forbearance, and my children so they may see improving quality in the world.

D.W.

For my wife, Cassie, and our two boys, who, at a cost to themselves, made this possible for me, and for the opportunity to improve how work delivers results—whatever you do, work heartily, as for the Lord and not for men.

F.P.

Table of Contents

List of Figures and Tables

Acknowledgments

The authors would like to thank the individuals who have contributed to this content: Anthony Agnone, Frank J. Alessi, Joan H. Alliger, Jonathan Andell, Yves A. Andre, Charles A. Aubrey, Robert B. Austenfeld, Dennis Beecroft, William J. Bennett, Roger Berger, Egbert F. Bhatty, Caroline J. Bolton, Dean Bottorff, Charles W. Bradshaw, John D. Breeze, Clayton C. Brewer, Clyde W. Brewer, F. X. Brown, Jack Campanella, Frank Caplan, T. J. Carlin, R. A. Cawsey, Ronald R. Cerzosimo, Alain M. Chauvel, R. Roy Cheeseman, Gary Cokins, Frank J. Corcoran, Lois Cowden, Phillip B. Crosby, Edward P. Czapor, Peter A. Daisley, B. G. Dale, Edgar W. Dawes, James Demetriou, William H. Denny, Richard K. Dobbins, Susanne Donovan, Grace L. Duffy, Ed Duke, Michael J. Dwyer, L. James Esterby, J. G. Fournier, James G. Fox, Harold L. Gilmore, William D. Goeller, Victor J. Goetz, William A. Golomski, Carl P. Green, Andrew F. Grimm, J. M. Groocock, Frank M. Gryna, Alvin Gunneson, T. Gurunatha, John T. Hagan, H. James Harrington, Charles P. Hollocker, Dan Houston, Zac Jarrard, Noriaki Kano, M. Karnebjer, John Bert Keats, William J. Kennedy, T. David Kiang, David D. Kildahl, David L. Kimble, April King, Yohei Koga, O. G. Kolacek, Herb Krasner, K. S. Krishnamoorthi, Robert C. Kroeger, Hugh E. Kroehling, Hitoshi Kume, William J. Latzko, John R. Lavery, Dennis D. Lee, Dean-Michael Lenane, Sanford Liebesman, Miguel Arenas Luega, Jd Marhevko, H. E. Mayben, Heather McCain, Peter Merrill, William N. Moore, Wayne J. Morse, August B. Mundel, Roderick Munro, Vyasaraj V. Murthy, Wiliam C. Noz, A. D. Oak, Duke Okes, William J. Ortwein, Bronislaw Oyrzanowski, Narendra S. Patel, Lawrence R. Petersek, J. J. Plumkett, Ronald L. Pollard, Kay Poston, Michael P. Quinn, Walter F. Raab, Govind Ramu, A. Raouf, J. P. Reames, Bradley F. Redding, R. C. Rhodes, Jim Robinson, Suhansa Rodchua, C. B. Rogers, Harold P. Roth, John M. Ryan, Frank Scanlon, William Schaeffer, Arthur M. Schneiderman, John Schottmiller, Lawrence J. Schrader, Doiran Shainin, Nick Shepherd, Walter Siff, D. Scott Sink, R. W. Stalcup, Robert G. Stenecker, Anthony R. Stephenson, Edward Sullivan, Earl T. Szymanski, Art Trepanier, Joseph J. Tsiakals, Thomas N. Tyson, K. Ullberg, Rudiger K. Vocht, H. Fred Walker, Paul A. Ware, Greg Watson, Tadeusz Wawak, James J. Wayne, Russell Westcott, Linda Westfall, A. Q. Whitton, W. C. Wilhelm, Ronald J. Williams, Noel Wilson, William O. Winchell, Jennifer Winchester, Martin W. Wirt, Paula W. Wright, A. H. Zaludova, James F. Zerfas, Debra A. Zimbler, Daniel J. Zrymiak.

The American Society for Quality and Quality Press would like to thank the Quality Press Peer Review Committee for its invaluable volunteer participation and contributions to this work. Without our volunteers' subject matter expertise, time, and passion for creating content, none of our efforts would be possible.

Committee Members
Scott A. Laman, Chair
Peter Pylipow, Vice Chair
Melvin Alexander
Ahmad Elshennawy
Gary Jing
Trevor Jordan
Jane Keathley
Jayet Moon
G.S. Sureshchandar
Theia Therurer
Mary McShane-Vaughn

Chapter 1

Quality Cost Concepts

The real value of a quality program is how it drives customer satisfaction and profit. While some organizations adopt quality programs to meet customer or regulatory demands, most managers know quality is about more than just avoiding issues.

WHAT IS QUALITY COST?

Reducing expenses or increasing revenue are two ways to improve profits. What is this topic that combines quality with costing? There is a common belief that higher quality comes at a higher cost and that reducing cost will necessarily result in a lower quality. While reducing material costs may lower product quality, consider the three definitions of quality.

1. Quality of features: A product with outstanding features and more features than competitors may be perceived as higher quality by customers.
2. Quality of reliability: A product that always does what it is meant to do and lasts a long time may be seen as higher quality by customers. Of course, more features may mean lower reliability.
3. Quality of process: Having a high-quality process will deliver both forms of product quality. In almost every instance, it will also lower the product's production cost by eliminating waste. Higher-quality knowledge, better monitoring, and better anticipation of customer requirements and supply chain risks help reduce operational costs while improving product and process quality.

A well-run quality program can reduce costs, improve product quality, minimize waste, and increase customer satisfaction, thereby boosting revenue.

THE BEGINNING: WHERE QUALITY AND COST COLLIDE

To talk about quality costs and the various concepts that lie behind them, it is helpful to begin with history, or "where did this all start." It began with Joseph M. Juran's first *Quality Control Handbook* (1951) in the first chapter: "The Economics of Quality." Juran wrote an analogy: "There's gold in the mine."

He meant that companies were wasting money on poor-quality production, even if they could fix it before the customer received the product. At that time, business economics was narrowly defined. Business economics focuses on expenses—labor, capital, and material. Quality was seen as unrelated to economics. A few papers expanded on this: W. J. Massers's 1957 article "The Quality Manager and Quality Costs," Harold Freeman's 1960 paper "How to Put Quality Costs to Use," and Armand Feigenbaum's classic book, *Total Quality Control,* were the first to classify quality costs according to the categories as they have been discussed for the last sixty years.

In 1961, less than a decade after Juran's work, the society then known as ASQC (American Society for Quality Control) established the Quality Costs Committee. (In 1997, the ASQC dropped the C and became ASQ, retaining the same functions.) The committee emphasized the magnitude and importance of product quality in a successful manufacturing business using dollar terms. The committee examined ways to measure these quality-related costs. The volunteer experts on this committee published the article "Quality Costs: What and How to Detail What Should Be Contained in a Quality Cost Program" in 1967. The committee also produced two volumes of papers in 1981 and 1989 under the title, *Quality Costs: Ideas and Applications,* and then the first two editions of this book, *Principles of Quality Costs,* in 1986 and 1990. By the third edition in 1999, definitions for the various categories and elements of quality costs, as we recognize them today, had been developed.

As organizational structures changed, the Quality Costs Committee joined the ASQ(C) Quality Management Division in 1983 and became one of its most active and productive committees. Valuable work was done, including developing the *Principles of Quality Costs.* As content reached a level of maturity, the Quality Costs Committee was discontinued in 1999. Although the pace of growth slowed, ASQ's Quality Management Division continues to support and update this body of knowledge to this day. As recently as 2024, ASQ (with support from the Quality Management Division) published "Technical Report 2," known as "Cost of Quality: Guidelines for Development, Implementation and Monitoring to Improve Quality and Performance ASQ/ANSI TR2:2024," to help ISO-9001 registered organizations apply quality costs.

While this field has matured and the rate of change has slowed, these concepts remain valuable. This book continues the work of the individuals who founded the concepts of quality costs and the committees that refined them so that all may apply them. Many firms have never used this approach, and they suffer for it.

Most research and measurement of quality cost has focused on manufacturing activities. As more work has shifted to office-based work, few have considered the office cost of quality (COQ). Even fewer have examined how poor-quality office processes lead to greater manufacturing waste. To remedy this oversight, this edition included research on the office COQ.

With the continued international popularity of the ISO series of standards, the concept of quality costs plays an essential role as a quality improvement tool and a measure of quality management. This edition includes updated references to ISO/DIS 9001:2025 (en), anticipated for release in 2026.

WHAT'S IN A NAME

Since 1999, the term *cost of poor quality* has sometimes been used as an equivalent to *quality costs*. It is not equivalent because it covers only a portion of the quality cost. In the Six Sigma literature, the term *cost of poor quality* is used almost exclusively. Think about the other piece: the cost of *good* quality. You realize that the term *cost of poor quality* in Six Sigma programs has reduced the emphasis on investments.

There's also the cost or investment in proactive quality that needs to be accounted for. In the last thirty years, continuous improvement programs such as lean, Six Sigma, kaizen, and others have worked hard to improve our processes and drive out waste. It is well recognized that these improvement programs need guidance. Improvements can't simply be applied wherever perceived issues appear, because many of those issues may occur significantly downstream from where the causes are developed. To manage a process effectively, managers need to identify where waste occurs and focus on proper monitoring and investments that drive the desired changes, regardless of the name of the improvement program.

An organization that effectively builds continuous improvement and sustains itself over the long term needs to understand where waste occurs, how much it costs, and how much it needs to invest in prevention and improvement programs. In essence, this is Cost of Quality (COQ).

The widespread adoption of process improvement approaches, such as Six Sigma and lean, has reduced quality costs to the cost of poor quality (COPQ). Such a focus on waste and downstream consequences may, however, have it backward. The use of quality costs across the organization, along with investments (prevention) and monitoring (appraisal) costs, enables the selection of appropriate improvement projects regardless of the improvement approach implemented.

This definition takes us beyond the actual expense of a Quality Department (managers and inspectors). It encompasses planning across the organization, involves a detailed study when aspects of the business go awry, and requires that we compare the cost of our mistakes to the investments we make in planning. If no errors are made in an organization, then no learning occurs. The only good purpose of a mistake is to teach. Our organizations cannot be trained if we are not monitoring our situations. In this sense, quality costs are a management tool for controlling and improving our work processes. Quality costs are not oriented toward counting defects or providing customer warranties. Quality costs look beyond these aspects, even though they can be counted as costs. Quality costs examine whether our planning, knowledge, and culture are sufficient to deliver the processes our organizations require.

It's not surprising that Six Sigma advocates use the term *COPQ* synonymously with *quality costs*, as their focus is on reducing variation, waste, and downstream consequences. Including investments in Six Sigma metrics would require including Six Sigma project expenses in the overall quality cost metric. When considered in this manner, it is clear that COQ offers a more robust and nuanced perspective on the costs than the COPQ example of Six

Sigma. This is not an encompassing analysis, but it may be a partial factor of Six Sigma's declining significance.

With that scope in mind, we also want the language of this book to be easy to teach and easy to use across functions. The following note explains the category terms we'll use throughout.

HIGH-LEVEL DEFINITIONS: A TERMINOLOGY SHIFT

With that scope established, if you've worked with quality costs before, you've likely seen the classic four categories: prevention, appraisal, internal failure, and external failure. Those terms carry decades of meaning for experienced practitioners. But for many readers—especially cross-functional partners and executives—they can sound technical, dated, or even accusatory.

In this edition, we keep the original intent of the four categories. Still, we describe them using a modern, business-friendly lexicon that translates more effectively across manufacturing, healthcare, service, and administrative environments (Table 1.1). This change is meant to make the system easier to communicate, teach, and sustain.

Table 1.1 Traditional vs. modern terminology.

Traditional	Modern
Prevention	Investment
Appraisal	Monitoring
Internal failure	Waste
External failure	Downstream consequences

A quick preview: *Investment* signals spending that builds future capability; *monitoring* captures inspection, auditing, and other detection activities across industries; *waste* avoids the "who failed?" framing and keeps attention on non-value-added effort; and *downstream consequences* acknowledges that external impacts often extend beyond warranty and returns. Chapter 2 provides the formal definitions, examples, and classification guidance you'll use in practice.

JURAN'S TRILOGY AS A NEW LENS FOR COQ

Joseph M. Juran described quality management as three connected activities: *quality planning*, *quality control*, and *quality improvement*. A quality cost system supports all three, but the work looks different depending on the purpose. In this book, the trilogy also helps guide organization of the material into distinct chapters.

After the definitions and measurement foundations in Chapters 1 and 2, Chapter 3 focuses on building a COQ program and setting it up for sustained use. That work naturally connects to planning because it requires making intentional choices about what to measure and how to prevent future losses. Chapter 4 then concentrates on how organizations use COQ to support control by making performance visible, detecting drift, and maintaining standards. Chapter 5 focuses on improvement by using cost information to identify the most significant opportunities, prioritize projects, and invest in future capability. These areas are tightly linked, and at first they can sound similar. The distinction is practical because each involves different questions, decisions, and behaviors. This structure also reflects a deliberate reconstitution of earlier-edition material, preserving what remains essential while making room for gaps, modern practice, and new areas of emphasis.

Organizations Are Unique, but COQ Always Works

Organizations may be unique, but the COQ framework remains effective because it is based on cause-and-effect thinking. It helps make quality-related losses visible enough to manage, reduce, and prevent. What varies from one organization to another is not the value of COQ but how it is applied, what is practical to measure, and where the most meaningful costs appear.

The first five chapters lay the groundwork for applying COQ in general terms. Many examples come from manufacturing, where COQ was first developed and refined and where categories and trade-offs are often easiest to see. However, the same principles apply to service and knowledge work settings. In these environments, "defects" frequently manifest as delays, errors, rework, unnecessary handoffs, and inconsistent customer experiences.

To help readers adapt COQ methods to their own context, Chapter 8 presents case studies from a variety of industries and organizational settings. Consolidating these examples into one chapter allows the book to explore specific topics in depth without requiring every reader to sort through examples that may not apply to their environment. This edition also separates two areas that require focused attention. Chapter 6 addresses supplier quality, since even vertically integrated organizations rely on suppliers. Chapter 7 explores software quality costs, where the project-based nature of software work reveals patterns that are useful across many types of organizations.

These chapters show that COQ is not a rigid formula. It is a flexible approach that adapts to different environments while helping organizations see and reduce the COPQ.

CONNECTING COQ WITH THE ORGANIZATION

COQ remains relevant because it connects quality improvement to decisions leaders already care about. But it only delivers results when it is implemented effectively and integrated into the organization's performance management. The remaining chapters focus on making that connection real and sustainable.

Chapter 9 places COQ within broader quality systems and widely used standards. It reinforces a key idea: COQ rarely succeeds as a stand-alone initiative. It gains traction when it aligns with how priorities are set, performance is reviewed, and corrective actions are carried out. In short, it must become part of the organization's regular way of working.

Chapter 10 tackles a critical issue. Whether COQ becomes a management capability or remains a measurement exercise depends on leadership support. It must be integrated into how the organization allocates time, attention, and resources. To support that integration, Appendix A has been updated and expanded to cover key financial concepts used throughout the book. These tools may be new to some, but they are essential for building credible business cases, analyzing trade-offs, interpreting results, and communicating effectively with decision-makers.

When quality professionals can frame technical insights in economic terms, it becomes easier for organizations to act earlier, fund the right efforts, and sustain improvements. The result is often greater impact, even without greater investment.

RESISTANCE: EVERYONE AGREES, FEW FOLLOW THROUGH

Adopting the COQ system can feel like stepping into a familiar conversation that everyone agrees is important but few know how to move forward. You're not alone if your organization recognizes that quality impacts the bottom line but struggles to connect that insight with real, measurable action. In fact, that's the norm.

According to the 2025 ASQE / IOE survey, overall, 85 percent of professionals agree that quality costs directly affect financial performance. Nearly half (54 percent) of respondents somewhat agree, while 34 percent completely agree. While that leaves plenty of room for improvement, resistance is rare: Only 15 percent expressed any doubt. Across industries, results are consistent, all reporting in the 80s, in line with the 85 percent mean. That kind of alignment is uncommon in most business initiatives, which is encouraging.

Of course, there are deviations from this mean. There is substantial variation between countries or regions. This will be important for practitioners trying to benchmark, but drawing simplistic conclusions from the results would be inappropriate, given shifting cultural values and standards that affect the results. Another vital divergence occurs even within organizations between executive and quality practitioners. In the survey, executives report significantly higher positive responses than nonexecutive quality practitioners. Again, like in culture, this may be a scale issue based on optimism or other value factors or on the necessity of the work. Executives have to know the relationship between actions and financial performance, but nonexecutives may also have greater clarity on the specifics of the COQ program.

Regardless of the causes, the long-suspected (at least by the authors) divergence between executive and nonexecutive responses is now quantitative, with a twenty-percentage-point spread of 95 to 75 percent. Points of differentiation are interesting and significant for analysis and learning. Still, the theme is consistency. Such alignment is uncommon in most business initiatives. It's encouraging.

But it's also telling that despite this consensus, many organizations still struggle to turn belief into behavior. And here's where things get familiar, maybe even a little frustrating: The top barriers reported aren't philosophical; they're practical and deeply human.

When asked for the single most significant barrier, responses clustered around leadership, technology, and communication. The top issue was lack of support or prioritization from senior leaders (16 percent), followed by technology limitations that constrain holistic analysis (15 percent). Breakdowns in communication from data collectors to decision-makers (13 percent) were close behind, along with incomplete or poor data (12 percent), ineffective reporting (12 percent), and gaps in employee skills or training (12 percent). Cross-functional cooperation challenges (11 percent) and inconsistent measurement methods (10 percent) rounded out the list. In other words, the pain isn't about whether to measure quality costs. It's about getting the right people, processes, and systems aligned, so the numbers are meaningful.

The real challenge isn't deciding whether to adopt COQ; it's figuring out how to make it work in organizations with siloed teams and outdated systems. If that sounds familiar, then continue reading.

The IoE survey data looks at the adoption journey through a practical lens, explaining why organizations commit to COQ. What motivates that commitment? And what often gets in the way. While we'll share data and patterns, the aim isn't to critique or prescribe but to help you see where your organization fits in this landscape and offer a path forward that feels achievable rather than overwhelming.

Implementing a COQ system is more than a technical task; it's a cultural shift. Success begins with clarity, shared purpose, and persistence. You're not starting from zero. You're starting from experience, instinct, and intent. This book is designed to walk alongside you, not just to inform but to support the process of turning insight into action.

A NEW BRANCH: OFFICE COST OF QUALITY

Over the last twenty years, employment in the United States has changed. Employment in the United States is shifting from factory work to office work. US government statistics indicate that between 1997 and 2011, factory jobs declined by 32 percent while office jobs increased by 39 percent. It is unlikely that this shift has reversed since 2011.

Office COQ waste shows up as lost productivity from inefficient meetings, distractions, email and communications churn, information search and software delays, health-related downtime, motivation/engagement issues, and rework from poor workflows or results.

Research on the office COQ, detailed in Chapter 8, identified several types of waste based on a literature review and interviews with consultants. The eight types of waste identified were meetings, distractions, emails and messaging, information storage / retrieval, health issues, motivation, workflow, and software / system delays.

WHAT GETS MEASURED GETS MANAGED

Quality metrics, such as the percentage of defective outgoing products or the fraction of defects found during inspection (or service failures), are often used to control quality within the quality management system. These metrics can provide strategic and operational goal orientation for the Quality Department and identify weaknesses in the quality situation. Still, such metrics have weaknesses in that they only show the extent to which targets for quality-related activities have been met. Furthermore, there is a time lag between implementing a quality initiative and the benefit payoff reflected in traditional cost accounting, which complicates isolating costs and benefits in quality.

Systematic reviews of articles from 1996 to 2025 consistently find that the benefits of using quality costs are evident beyond those of a general continuous improvement approach. These fall into broad categories, including communication, motivation, improved data-based problem-solving, and enhanced leadership.

The benefits of the COQ metric can be examined as internal (direct or indirect) or external.

Internal Benefits—Direct

- Reduce production expenses by providing a qualitative measure of waste
- Enhance profit by reducing waste in focused areas, avoiding broad cuts in allocation that might reduce capacity
- Gain a higher return on quality improvements (ROQI), which indicates that quality improvement efforts have been effective

These direct benefits can be substantial, but the variety of indirect benefits is wide.

Internal Benefits—Indirect

- Compare process improvements and identify the most cost-effective options
- Compare the success of projects or initiatives
- Use quality cost data for motivational purposes, portraying results in numbers to which members of a department directly contribute (such cost transparency engages employees in a team effort to make real improvements)
- Find quality improvement candidates by revealing areas with higher quality costs, whether they are finding opportunities for improvement or optimization

- Provide a way to fulfill a stakeholder's requirements, if they are seeking a better grasp of quality
- Identify soft targets for improvement efforts, along with either value-adding activities or non-value-adding activities
- Allow budgets based on expected returns and justify quality management—in general, a basis for budgeting the quality operation
- Find an appropriate trade-off between the levels of conformance and nonconformance costs

As significant as these internal benefits are, we can add some external benefits as well.

External Benefits

- Reducing internal defects shipped to customers increases customer satisfaction, which in turn drives greater customer loyalty and organizational revenue
- Increased customer satisfaction will also increase purchase frequency and/or quantities
- As satisfied customers recommend the product or service to others, the resulting communication will benefit the organization's reputation

As no organization exists without customers, customer focus is one of the most critical aspects of a quality assurance system. So the real value of any quality program lies in its ability to drive customer satisfaction, leading to increased revenue and improved profits.

It follows that one primary benefit of a COQ program is to enhance customer satisfaction and loyalty. While customer focus in business-to-business differs from that in business-to-consumer organizations, a COQ program must recognize downstream consequences arising from internal issues.

FACTORS HINDERING THE ADOPTION OF QUALITY COSTS IDENTIFIED IN THE LITERATURE

Academic research takes a different approach to roadblocks than practitioner-focused studies, such as those from IOE. Together, they give a fuller picture. For more on the gap between management and quality perspectives, see Chapter 10. In S. Rodchua's 2005 thesis, a survey asked about problems in implementing quality cost programs. The main issues were a lack of leadership involvement, an embedded culture, difficulty obtaining reliable data, a lack of Accounting Department involvement, and limited knowledge of quality costs. Other issues included focusing solely on surface-level profitability, weak information technology (IT) infrastructure, and the belief that technology alone will solve problems. These are shown in Table 1.2.

Table 1.2 Excuses for not measuring quality cost.

Excuse category	Detailed reasons from the study
Management	Lack of top management support
	Organizational culture is historically resistant to change
	Human resource management is not engaged to motivate members of an organization, has no strategy for employee development, and has no plan for customer satisfaction
	Using a traditional adversarial relationship for supply chain management, limiting teamwork among supply chain partners
	No belief in management by fact
Strategy and culture	A belief that quality metrics are for quality people, and cost metrics are for accounting people
	Past implementations have lacked crucial contributions of employees and shown inconsistency and incoherence
	Current subject matter experts do not see the implementation of new management approaches as valuable
	Lack of cooperation between functional areas
	The organization is entirely outward focused on customers and does not connect internal metrics with customer satisfaction
Magic beliefs	The organization focuses on IT to improve products and services, considering employees as secondary to success
Profitability	Outstanding financial performance results in reduced attention to internal waste
Knowledge	Considering quality and cost as opposites, a zero-sum game
	Lack of knowledge of the COQ
	Confusion about the terminology of the COQ
	Not having access to data

THE BARRIERS BETWEEN BELIEF AND BEHAVIOR

The 2005 survey is a good starting point, but we can add to it. Consideration of hindering factors is essential for smooth transitions and the adoption of changes. The following are additional factors that, in our experience, may be present:

Missing Leadership Involvement

- Lack of cooperation from the senior leadership team and management's attitude

- Issues of not attracting engineer talent, multitasking, and poor project planning
- Not getting senior management to understand and accept the value of looking at this information and taking positive steps toward improvement
- Staff intransigence
- Reluctant people, lack of commitment
- Poor acknowledgment of quality cost efforts for quality improvement
- Too low a priority level for COQ
- Not driving it to all levels
- Lack of upper management support and lack of middle management involvement and ownership of their piece of the pie
- The management staff does not want to spend the money to set up a sound system or take the time to learn about the process
- Missing sponsorship, understanding, and not linking to business goals
- Not getting all key areas on board and treating COQ as a "quality" activity only

Culture

- Enabling a culture that favors prevention over correction
- Management is hiding expenses to make performance appear more effective
- Fear of head count reduction if costs go down
- Unless there's a high cost, no one pays any attention
- Lack of discipline in reviewing and reacting to negative trends
- People don't want to share the problems and the costs associated with them
- Not reacting when data shows a problem
- Resistance of employees to changes, particularly in troubleshooting opportunities for improvement

Gaining Data

- Not gathering COQ data
- Cost and quality systems generally do not develop with this type of data collection, with costs reporting in mind; data is abundant; information is rare
- Missing collection of the data
- Deliberately hiding costs
- Missing consistency from plant to plant
- Not tracking precise data
- Rework labor information is wholly or partially estimated

- Not getting real-time or accurate scrap data; in an industry that is a high-volume, low-value industry, scrap rates tend to be higher, but in a relatively lower-dollar-value raw material business, scrap costs may be ignored
- Not obtaining accurate costs
- Not keeping the collection of internal failure costs simple at first
- Inaccurate scrap data from the shop floor and/or erroneous data entry
- Pursuing perfection in the calculation and costing at the expense of sound judgment

Missing Accounting Involvement

- No involvement of the controller, so not using actual costs versus estimated costs
- Stopping when a system has defined stable categories for use as a barometer in key areas versus trying to capture significant quality costs company-wide
- Not certifying the organization's members in accounting to enhance credibility
- When accounting is not involved

Lack of Knowledge of Quality Costs

- Not learning quality, costs, scope, and classifications
- Lack of understanding of the concept and impact of the COQ by those in decision-making positions
- The company is unaware of quality costs

Customer challenges

- Short product development expectations from customers that don't give adequate information to design properly

BEFORE THE NUMBERS: CORE IDEAS UNDERPINNING COQ

Now that we've discussed benefits and barriers, we need to look at the core principles in quality cost measures. There are three central concepts behind quality costs and their measurement.

1. *Quality costs are a tool for managing processes and making improvements.* Knowing where waste exists targets improvement efforts; proper investment brings about change; and monitoring serves two purposes: informing management of current conditions and ensuring that improvements are maintained. Downstream effects are similar to waste measurement; they reveal weak points in both dollar value and intangible areas, helping ensure that the organization's broader goals are achieved effectively.

Figure 1.1 Hidden costs of quality and the multiplier effect.

2. *All costs related to quality do not need to be identified, since many are hidden and difficult to measure.* These hidden costs often exceed those that are easily seen and measured. A small visible bit of waste may link to an unseen set of costs, similar to the way 90 percent of an iceberg is underwater, as in Figure 1.1. For example, one COQ is a missed shipment to a major customer. The obvious costs here are initiating a production run just for that customer, paying for expedited shipment to that customer, and other easily measured costs. If you think about it, setting up a special production run for a customer requires additional management time and staff effort, detracting from other valuable work. This extra effort could be considered a hidden cost and hard to measure, as it will likely be only a small portion of the activities those individuals typically perform. It is a cost, and it would not have occurred had the shipment not been missed.

   If you measure quality costs based on your internal systems or the nature of your business, and you invest properly in control methodologies, you will not only reduce the quality costs you measure but also the hidden costs. In this sense, measuring quality costs differs from other cost accounting programs. For most organizations, cost accounting programs measure all costs and attribute them to production activities to provide a complete picture of the organization's expenses and profitability. As a management tool, quality costs do not need to measure all expenses, only the significant ones.

3. *Greater investment in prevention results in lower waste-related spending, and prevention is always cheaper than waste.* Building on the second concept, investing more will reduce waste and downstream consequences. Over time, improved processes reduce the need for monitoring, though during

changes, there may be a temporary need for additional monitoring. In his 1979 book, *Quality Is Free,* Philip Crosby clearly laid out these ideas. He was fighting the perception that producing a quality product was necessarily more expensive for an organization. Crosby made the point that improving quality was not necessarily about spending more on materials or labor; a higher-quality, defect-free product drove higher revenues and lower expenses far more than the investment required to improve quality. Therefore, higher quality was "free."

The core concept is that the third concept is sometimes called the 1:10:100 Rule. Simply put, the earlier you can prevent a problem in your process, the less expensive it will be to fix. Careful attention as you progress through the stages of development and production helps catch issues early. Starting with design and continuing through planning, setup, production, and distribution, the cost of a mistake increases by a factor of ten at each significant stage. What might cost you $1 to fix in the early design stage would cost you $10 as you proceed with planning and involve a larger team. If the problem is not discovered until you reach your process setup, it can cost you $100. If you were in production, the mistake could cost you $1,000, and once you've made your mistake, the cost will easily exceed $10,000. While these actual dollar amounts may not be precise, their proportions are valid. This places the burden of effort on early detection and prevention, thereby emphasizing our first category of quality cost investment. See Figure 1.2.

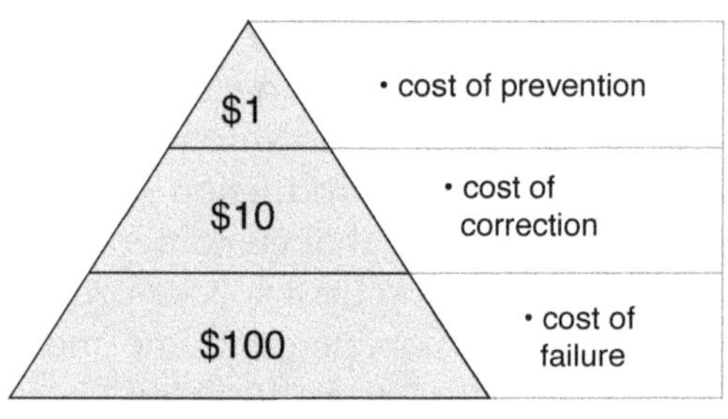

Figure 1.2 The 1:10:100 rule in action.

COQ IS NOT ACCOUNTING, YET ACCOUNTING HELPS

In cost accounting, the most important thing is to account for all material costs. Any expenses that are not counted or omitted can pose a risk to the organization's reporting. In the worst case, uncounted costs may cause your revenues to fall short of your expenses. In other words, you may be losing money.

We listed our four categories of quality costs: investment, monitoring, waste, and downstream consequences. There's a fifth category of business expense that is not part of quality costs and can best be described as the cost of doing it right the first time. Labor and materials are expended to make a good product and deliver it to the right customers at the right time: This is the cost of doing business, not a COQ.

The measure of quality costs is an abstract of the full cost accounting within an organization. You don't have to measure all of your quality costs; quality costing is a management tool that places it in a different category from broader-based total cost accounting.

For many organizations, there's a firmly held belief that they run a tight ship. In other words, they believe that they generate very little waste and recognize they are near maximally profitable. Being profitable simply means that your revenue exceeds your expenses, but it does not mean your expenses do not include waste. Firms like this believe that their present cost accounting system is sufficient to identify all areas requiring management attention. Unfortunately, accounting systems were not designed to demonstrate the impact of quality on performance or how it changes overall operating costs. This belief can lead organizations to think that implementing a quality cost program will require extensive changes to their accounting systems and additional staff that may not be needed.

In every organization we have seen implement a quality cost program, the leadership was surprised by the size of the waste cost. It can be challenging to convince leadership in such an organization to begin a program to measure quality costs. Their shareholders are not asking for it; nor are their customers. In terms of measured quality costs, they might see the quality cost program as a source of competitive advantage if they adopted it before their competitors; it can give them an advantage.

It's difficult to measure quality costs across industries because most firms treat the cost of their mistakes as an internal matter and do not share it with outsiders, especially customers. Also, each organization has its own view of where its waste occurs and what causes it. The COQ measure is usually tailored to the organization and is difficult to compare with others.

Launching a quality cost initiative can be daunting, but experience across a wide range of industries shows it can be a meaningful and worthwhile endeavor. The remainder of this book expounds on and clarifies the definitions of the four categories, provides practical advice and knowledge to help launch a program, and ensures that the program first effectively controls quality and then leverages that control to improve overall quality. The text also examines specific applications, such as supplier quality or office waste, that will help in that context and provide illustrative nuance that one may find widely applicable. Many hours of research and practical experience have expanded on what our COQ pioneers have built, so we hope you enjoy it and find it insightful, practical, and valuable.

KEY TAKEAWAYS

- Quality costs are a management tool, not a complete accounting system: Intentional focus should be on the most significant quality-related expenses to guide decisions, not on capturing every cost.

- Higher quality does not inherently mean higher cost. When managed through quality costs, wise investment in quality (upstream) reduces total cost by lowering waste and downstream consequences.
- Modern terminology—*investment, monitoring, waste,* and *downstream consequences*—better reflects how organizations think about expenses today than the traditional *prevention/appraisal/failure* labels.
- Quality costs extend far beyond the Quality Department and the shop floor. Office processes, information flow, and decision-making can generate substantial, often hidden, quality-related costs.
- Many quality-related costs are hidden or challenging to measure (e.g., extra coordination, special runs, expediting). Measuring the visible portion and acting on it also reduce these hidden costs.
- Investing earlier in the life cycle is far cheaper than fixing issues later; the 1:10:100 Rule highlights how prevention during design and planning dramatically reduces costs in production and the field.
- There is a distinct "cost of doing it right the first time" that is not a quality cost; quality costs focus on the extra expenses associated with preventing, detecting, and correcting problems.
- Organizations often underestimate the size of their waste. When quality cost programs are implemented, leaders are frequently surprised by how large waste and its downstream consequences truly are, revealing opportunities for improvement and competitive advantage.

Chapter 2

Quality Cost System Definitions

Organizations invest significant resources—labor, materials, and time—to ensure the quality of their products and services. Managing these investments effectively requires a structured approach that outlines how funds are planned and used. A quality cost system provides this structure by categorizing expenses to reveal opportunities for improvement and support informed decision-making.

Traditionally, quality-related costs have been grouped into four categories: prevention, appraisal, internal failure, and external failure.

For clarity, this book adopts updated terminology (as shown in Chapter 1, repeated for emphasis in Table 2.1).

Table 2.1 Modern terms.

Traditional term	Modern term
Prevention	Investment
Appraisal	Monitoring
Internal failure	Waste
External failure	Downstream consequences

These updated terms make each category's purpose clearer. For example, *investment* highlights spending to prevent problems, and *downstream consequences* points to the effects of quality failures after delivery.

In addition to these four categories, most organizations incur a fifth type of expense: the cost of doing things right the first time. While this represents the bulk of total spending, it is not considered part of quality costs because it does not help measure the effectiveness of a quality system.

Understanding these categories—and how they relate to such concepts as the *cost of conformance* and the *cost of nonconformance*—is essential for building a robust quality cost framework. In the sections that follow, we define each category in detail, explain its role in quality management, and provide practical examples to guide implementation.

RENAME THE BUCKETS, IMPROVE THE CONVERSATION

There are actually two sides to quality costs: the cost of poor quality (COPQ) and the cost of good quality. These may also be defined by the terms *cost of nonconformance* and *cost of conformance.* The four categories outlined above don't align precisely on these lines. While investment expense is clearly included in the cost of conformance, and waste and downstream consequences expense categories are clearly included in the cost of nonconformance, monitoring expense may be considered either a cost of conformance or a cost of nonconformance. Some form of monitoring will always be necessary, and monitoring expenses are not totaled in either category. Monitoring expenses quantifies how you're doing, an essential input to decision-making. Monitoring costs may be split between the cost of good quality and the COPQ.

Let us define the four categories more thoroughly.

Investment

The costs of all activities specifically designed to prevent poor quality in products or services. This generally includes all quality planning and many aspects of project management. Examples include the costs of customer / user perception surveys / clinics, new product reviews, contract / document reviews, quality control planning, supplier capability surveys, process capability evaluations, quality improvement team meetings, quality improvement projects, and quality education and training.

Monitoring

Costs associated with measuring, evaluating, or auditing products or services to assure conformance to quality standards and performance requirements. Examples include the costs of incoming and source inspections / tests; product, process, or service audits; specific observations of personnel (including operations personnel) to ensure quality performance; acquisition costs of measurement equipment; calibration of measuring and test equipment; laboratory support; and the costs of associated supplies and materials.

Waste

Costs resulting from products or services that do not conform to requirements or customer/user needs before delivery of the product or completion of the service for customers/users. Examples include costs of scrap, rework, reinspection, retesting, material review, disposition of purchased material rejects, downgrading of product, issuing credit for dissatisfied service customers, below-target yields or efficiency, and inventory scrap due to design changes. For office waste, poor results, and workflow issues, the associated operational costs may also be added.

Downstream Consequences

Costs incurred after delivery or shipment of the product or after (or during) furnishing a service to the customer. Examples include the costs of processing customer complaints, customer returns, warranty claims, and product recalls. Liability costs may be added if significant or if specific events increase them. Some downstream consequences may occur well after the initial delivery or completion of work, and not all consequences may be measurable. For example, damaged customer or user goodwill is a cost because goodwill is listed as an asset on many formal accounting systems.

A Place for Everything and Everything in Its Place

Mislabeling costs can lead to poor decisions. For example, if you call incoming material inspection an investment, you might overlook the need to fix the root causes of defects. If you treat startup material and labor as just the cost of doing business, instead of waste, you may miss chances to cut startup waste.

Detailed breakdowns of the elements in these four quality cost categories are provided in Appendix B. These examples were developed over many decades of teamwork by dozens of quality professionals and have been updated for this edition. These elements will vary by industry. Quality cost elements in healthcare, for example, will differ significantly from those in manufacturing. Because of the extent of these differences and the many sectors involved (e.g., banking, insurance, hospitality), no attempt will be made to provide complete lists of these elements by industry. Each organization needs to define its own terms. Appendix B can serve as a guide and starting point for this activity.

NORMALIZE THE NOISE: WHEN WORKLOAD CHANGES, KEEP THE SIGNAL CLEAN

In developing the details of an individual organization's quality cost system, the quality manager and other management must collaborate to integrate their various sources of knowledge into a single system. Because costs may be incurred by any department, function, or cost center, a customized internal quality cost procedure is required. This procedure describes the data sources to be reported or extracted from the account ledgers, using existing account, department, and cost center codes. It will explain how any required estimates are to be prepared and where to count associated labor benefits, allocated costs, and labor burdens. It will provide the measurement bases against which quality costs may be compared.

Actual dollars expended are usually a good indicator for determining where quality improvement projects will have the greatest impact on profit and where corrective action should be taken. Still, unless the amount of work performed is relatively constant, this will not provide a clear indication of quality cost improvement trends.

Remember, a key value of a quality cost system is identifying opportunities for improvement and measuring that improvement over time. Since the volume of business, in total or in any particular product or service line, will vary over time, improvements in the cost of quality (COQ) can best be measured as a percent of, or in relation to, an appropriate base or denominator. Total quality cost relative to an applicable base yields an index that can be plotted and periodically analyzed against prior indexes. The base used should be representative of and sensitive to fluctuations in business activity.

For long-range analyses, net sales are the most commonly used base for presentations to top management (Figure 2.1). For example, a goal for the total COQ may be to reduce it from 9 percent of sales to 8 percent over a given business year. While this measurement may be necessary for top-level strategic planning, it would not be practical for measuring the day-to-day, week-to-week, month-to-month efforts of lower-level practitioners responsible for making it happen. There may be a considerable time lag between incurring quality-related costs and booking net sales. Timing variations between the two may put them out of sync, leading to varying ratios and excessive noise in the ratio metric. Consequently, net sales are the best basis for high-level, long-range analysis and tracking.

For example, in aircraft manufacturing, failing to ship even one aircraft during the quality cost reporting period could significantly impact sales for that period. Sales for the period would dramatically drop, thereby raising the quality cost index, even if quality performance had not changed. Going one step further, the delayed sale of that aircraft in the following period might inflate that period's sales figures, thereby creating a significant, misleading quality improvement trend relative to the previous period.

If your sales vary significantly from period to period, net sales are not a sound short-term basis for comparison. Over the long run, though, these ups and downs usually balance out, so net sales are a good metric for long-term tracking. Rolling averages can help smooth out short-term swings in any base. They do add some lag and can hide sharp changes, but they allow you to avoid overreacting to random fluctuations.

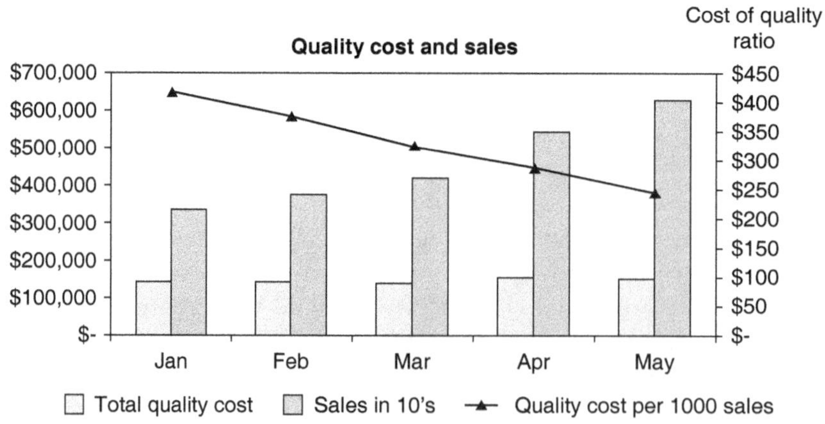

Figure 2.1 Using sales as a base for comparison.

Short-range bases should be closely related in time and location to the quality costs as they are incurred and reported. They should report the COQ as a percentage of the amount of work performed at a measurement location. For short-range use, appropriate bases for quality costs are best determined by reviewing volume metrics already in use in work areas and understood by the people who will use quality costs. The best bases are those that are already key production measures. Typical examples include overall operating costs, total or direct labor costs, value-added costs, and the actual average cost of the delivered product or service, as shown in Figure 2.2. The core idea is to use a meaningful, well-known base for the amount of business activity in each area where quality cost measurements are taken.

Cost of goods sold is often used, as it represents a *cost* basis. Sales are reported on a *revenue* basis, with all the adjustments between cost and revenue. For example, sales may be delayed significantly after production costs; the profit mix in sales can vary in ways that obscure the COQ; and competitive pricing may create temporary changes that are inappropriate for an internal metric, such as quality cost. For this reason alone, lost revenue due to quality issues should be excluded from quality costs; we cover other reasons not to count lost revenue in Chapter 3.

Figure 2.2 Using wholesale revenue as a base for comparison.

For effective use of a quality cost system, it may be preferable to use multiple bases. As mentioned above, in long-range planning, total quality costs as a percent of net sales are often used. For current, ongoing applications, the following examples are typical indexes that help avoid allowing volume variations to disrupt the COQ metric:

- internal waste cost as a percent of total production/service costs,
- downstream consequences divided by shipment volume,

- procurement monitoring costs as a percent of total purchased material costs,
- operations monitoring costs as a percent of total production/service costs, and
- total quality costs as a percent of production/service costs.

There is no limit to the number of bases or indexes or to the level of detail that an effective quality cost system may have. More danger lies in oversimplification—such as using a single base for all purposes. There is no single perfect base. Each base can be misleading if used alone. This can easily lead to confusion and disinterest. It's vital to the success of quality cost that the bases for individual progress measurements not appear unnatural to the functional area. Instead, they should be seen as complementary (e.g., rework costs as a percent of area labor costs). They could also be used to provide indexes that have shock value, simply to get the corrective action juices flowing (e.g., "Hey, did you know that for every dollar expended in your area, fifty cents is the COPQ?").

To help in the selection process, consider the following types of commonly available bases:

- a labor base (such as total labor, direct labor, or applied labor),
- a cost base (such as shop cost, operating cost, or total material and labor),
- a sales base (such as net sales billed or sales value of finished goods or services), and
- a unit base (such as the number of units produced, the number of services performed, or the volume of output, as in Figure 2.3).

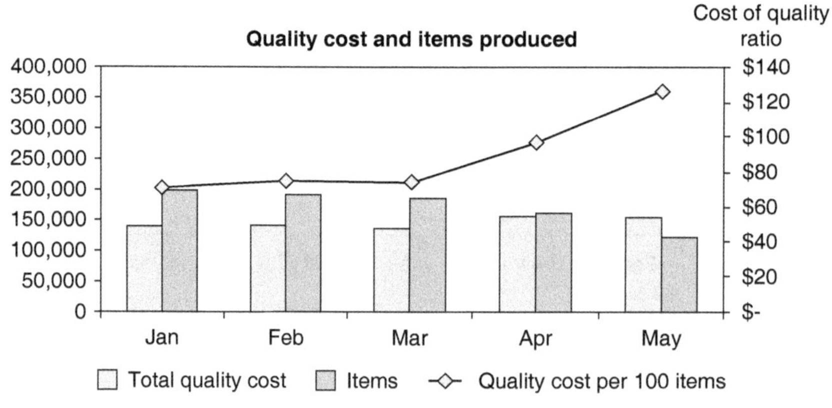

Figure 2.3 Using items produced as a base for comparison.

Additional factors can influence the application of these bases:

- *Sensitivity to increases and decreases in production/service schedules.* Most manufacturing and service operations have an efficiency peak. Additions and subtractions to the workforce, equipment maintenance, and the use of backup suppliers may influence both quality costs and built-in prime costs. If the influence is substantial, an attempt should be made to quantify these factors and recommend changes to minimize adverse effects. Successful companies use many techniques to overcome such adverse influences, including flexible manufacturing systems, intensive formal training, and lean quality programs with suppliers.

- *Automation.* With productivity and quality as goals in business competition, successful companies have turned to robotics, automation, and artificial intelligence to reduce direct and indirect costs. Here again, the effects on such ratios as scrap, rework, or monitoring versus direct labor costs may be substantial. Obviously, the application of quality cost principles dictates that we be able to measure the (presumed) favorable influence of automation on monitoring or waste costs and, more broadly, on a business's ability to influence customer perceptions and experience of quality.
- *Seasonal product sales.* Some companies have high seasonal sales. Downstream consequence costs, such as customer complaints, may be seasonally grouped, and quality costs adjusted accordingly. A four-quarter moving average of the ratio of downstream consequence costs to net sales billed is an appropriate technique in these circumstances.
- *Oversensitivity to material price fluctuations.* The law of supply and demand still prevails, and raw material costs may fluctuate widely. If waste or monitoring ratios vary due to the application of prime costs, this may have an unnatural, dramatic effect. Using direct labor (rather than prime costs) may be appropriate.

THE MONEY-LANGUAGE TOOLS THAT MAKE COQ WORK

To understand the strategic value of quality initiatives, it helps to view them through a financial lens using a small set of tools common in management decision-making. These ideas are introduced briefly and used selectively throughout the text, but they are essential for linking quality efforts to results leaders care about. Additional detail is provided in Appendix A.

Time value of money (TVM): This principle holds that the value of money changes over time. Investments in quality today—such as process improvements or training—yield benefits in the future. Recognizing TVM helps prioritize quality efforts that provide greater returns sooner rather than later.

Present value (PV): PV helps us assess how much future cost savings or revenue from a quality initiative are worth today. It's essential to compare up-front costs with long-term gains to make a clearer judgment on whether a quality investment makes financial sense.

Future value (FV): FV is used to forecast how today's investment in quality will grow over time. It helps frame discussions with financial stakeholders who may want to see the projected economic benefits of today's decisions.

Net present value (NPV): A key tool for evaluating quality projects, NPV compares expected cash inflows and outflows over time. A positive NPV indicates the initiative should increase the company's value.

Internal rate of return (IRR): This is the rate at which a project breaks even in terms of discounted cash flows. It provides a valuable benchmark for comparing competing quality projects and aligning them with corporate investment expectations.

Payback period: A simple measure of how long it takes to recover an investment. While it lacks the depth of TVM-based tools, it is still helpful for understanding risk and liquidity, especially for short-term quality efforts.

These financial concepts ensure that quality-related decisions are not only technically sound but also financially justified. They help link process excellence to enterprise performance, an increasingly critical expectation of modern quality management.

CONCLUSION

Quality costs are not just accounting categories. They are a practical way to see how well the organization converts effort into results. When buckets are clearly defined and labeled in plain language, people stop debating semantics and start fixing problems. Just as important, the measures must remain honest. If workload, seasonality, automation, or sales timing distort the ratios, you end up reacting to noise instead of improving performance. A well-built COQ system does more than report costs. It creates a shared, money-based view of where to invest and monitor and where waste and downstream consequences quietly drain value.

KEY TAKEAWAYS

- *A quality cost system turns spending into insight.* It organizes quality-related expenses to enable leaders to prioritize improvement.
- *Clear labels improve decisions.* Investment, monitoring, waste, and downstream consequences make intent easier to understand and harder to misclassify.
- *Monitoring is necessary but not always good.* It can support conformance or reveal nonconformance; treat it accordingly.
- *Mislabeling costs hides root causes.* Avoiding calling an inspection an "investment" or treating startup losses as a cost of good quality can help prevent waste.
- *Trends require the right denominator.* Net sales can work for long-range views, but short-range control needs operational bases tied to the work.
- *Use more than one base.* Multiple indexes reduce distortion and make the metric meaningful to the layfolk and domain experts.
- *Finance tools make COQ actionable.* TVM, NPV, IRR, and payback help connect quality decisions to business value.

Chapter 3

Build a COQ Program That Leadership Funds and Teams Use

Like many good things in life, a quality cost program doesn't happen spontaneously—it requires effort, advocacy, and leadership within the company. While the quality manager or senior quality representative often takes on this role, any committed individual can lead the initiative. Several key attributes are necessary for success:

- a solid understanding of quality cost systems,
- an apparent belief in their value and application within the company,
- a willingness to advocate for and lead the program, and
- a role that provides the opportunity to address challenges effectively.

It is also beneficial if the program's champion has access to and an understanding of financial records, such as the general ledger and profit and loss statements, to ensure direct financial alignment.

To help practitioners move from intent to execution, this book recommends the LEADER (lead, estimate, align, design, embed, run) Framework as a simple, practical path for launching and scaling a quality cost program. The framework doesn't include every possible detail. Instead, it is an abridged approach to organizing proven practices into a sequence that reduces uncertainty and apprehension while keeping the effort moving forward.

Too many COQ initiatives stall not because the concept is flawed but because teams struggle with where to start, how much rigor is "enough," or how to maintain momentum once early enthusiasm fades. The LEADER Framework addresses these common friction points by providing a straightforward structure that helps practitioners avoid analysis paralysis, prioritize decisions, and keep the project started and moving.

The framework is intentionally ordered. Each step builds on the one before it, creating a steady progression from sponsorship to scale:

- Lead with a champion and an accounting colead.
  Establish ownership and financial credibility from the start.

- Estimate the opportunity using the least arguable values. Focus on conservative, defensible numbers to build trust early.
- Align leadership on what "credible" means and what decisions COQ must support. Ensure agreement on purpose before expanding the scope.
- Design a low-risk, high-signal pilot with clear selection criteria. Prove value quickly without overextending the organization.
- Embed procedures, definitions, and responsibilities, so the program survives scale. Move from a project mindset to an operating discipline.
- Run the cadence, trends, reporting, and phased expansion. Create rhythm, visibility, and sustained improvement over time.

This chapter examines these steps and expands on them with many additional details. For readers who want a quick reference while applying the concepts, a concise LEADER Framework tool is provided at the end of the chapter for easy reuse in real-world settings.

The goal is simple and essential: Give practitioners a clear path forward, reduce hesitation, and replace uncertainty with disciplined progress while also providing a comprehensive review of implementation best practices for those who want to wade a little deeper.

DOING THE HOMEWORK BEFORE ASKING FOR SUPPORT

The first step in gaining management's attention—the most valuable resource in any organization—is to present a compelling case with both data and narrative. An initial review of financial data should provide a broad understanding of existing quality costs. Key cost areas are often already documented in previous analyses, such as scrap reports or quality system management reviews. If exact figures are unavailable, reasonable estimates can still provide valuable insights.

This initial assessment need not capture every detail; it should focus on significant quality costs that are readily available or easily estimated. If these figures represent a considerable portion of overall expenses, they will be sufficient to engage management and justify further investigation. It is also important to identify significant hidden quality costs, such as substantial rework expenses buried within general operating accounts.

Estimate your quality costs. Then highlight the most significant improvement opportunities. These insights should be enough to capture management's interest, and the remainder of this book explores how to transform that opportunity into a sustainable, structured program. Often, initial quality cost estimates exceed 15 percent of sales—even in profitable companies. Some manufacturing firms with well-established quality improvement programs have reduced total quality costs to 2–4 percent of sales, with those savings contributing directly to pretax profits.

Getting to Yes on COQ

The next step is to determine whether management is willing to support a quality cost program. This decision involves a strategic allocation of resources. While many quality professionals understand the long-term benefits of COQ management, executives and senior managers often have broader considerations. Their decisions must balance factors beyond profitability, including short-term cash flow (even profitable companies can fail if cash flow is mismanaged), investments in higher-growth markets or divisions, and trade-offs between current operational effectiveness and future innovation.

The role of quality leaders is to present the strongest possible case for COQ while trusting that company leadership will make informed decisions. One of the most significant advantages of a quality cost program—something other quality initiatives should learn from—is that it speaks the language of executives: financial impact. However, no quality cost program can succeed if driven only from the bottom or middle of an organization. Like all performance improvement efforts, it requires the support and commitment of top management.

Once it is clear that senior leadership is open to the idea, the next step is to craft a compelling pitch that secures their full backing. In addition to the broad financial overview already developed, this may require a more detailed secondary analysis. This analysis should focus on a specific operational area and demonstrate how quality costs can be accurately identified and reduced through corrective actions. A targeted secondary review helps leadership understand the full cost-reduction and quality-improvement cycle. If that aids acceptance, resourcing, and execution of the analysis, the secondary review may be conducted within another operational framework, such as a preexisting one, such as lean or Six Sigma's Define-Measure-Analyze-Improve-Control (DMAIC).

Engaging management in the process is key. Instead of presenting a theoretical argument to executives, involve them in defining the follow-up study. Ask them what they need to see to make a decision. When leaders are actively engaged, they are more likely to champion the program and share its potential with peers. By the time the final approval meeting takes place, the key decision-makers should already be convinced.

Stakeholder management is key, and the Stakeholder Analysis Guide can help. The tool detailed at the end of this chapter is a customizable guide that specifies who to include, questions to ask, and a table to visualize and manage progress.

If direct executive input is unavailable, another approach is the MOUND ("middle out, up, and down") strategy. When middle management drives improvement efforts that align with financial objectives and company goals, their successes naturally gain visibility. Senior leaders will review the results and share them with their peers, who can then initiate similar initiatives within their own departments.

BUILD THE RIGHT TEAM, NOT JUST "QUALITY"

The possibility of broad interest in a COQ measure will mean input from a diverse set of functional areas to make sure any volume adjustments to the dollar COQ are sensible to the functional areas. This will drive a range of cross-functional team makeup. In addition to operations, quality, and accounting, other groups, such as engineering, logistics, supply chain, and other support areas of operations, are needed to reduce quality costs. All should be a part of the team.

The impetus for a quality cost program typically comes from the Quality Department itself. Quality management (including managers, directors, and vice presidents) is usually interested in how to communicate with top leadership about the impacts of poor quality. They are also looking to find ways to justify investments to improve quality and customer satisfaction. The focus of most quality professionals is on statistics, measurement, inspection, and specification; a cost accounting group typically performs tracking expenses. Involving key personnel from the cost accounting function in the quality cost program will help bridge the gap between current accounting data and the program's needs. The individuals selected from accounting to participate need to know how the books should be kept and, more importantly, exactly how they are kept. Having accounting personnel help build the COQ metrics will increase top management's trust in the reporting. After all, the accounting function is where leaders look to obtain accurate cost information.

This does not negate the fact that a quality cost program is best served when it is not reliant solely on accounting personnel but is guided in partnership with senior quality leadership. Under generally accepted accounting principles, cost reports should reflect the total expenses for given functional areas. When we examine COQ metrics, we often find that they draw on expense data from different functional areas. In COQ, some expenses may need to be captured not in the area that does the work, as is typical in accounting, but in the region responsible for the process that drives results.

For example, if incoming quality monitoring is performed by operations personnel and supplier selection is made by separate procurement personnel, including monitoring costs in a procurement COQ report may encourage better supplier selection. If the procurement group cannot find high-quality suppliers, costs for this monitoring effort in the production area will increase. By reporting this monitoring expense to procurement leadership, they can be incentivized to modify how suppliers are engaged and selected. Because this type of reporting activity runs counter to generally accepted accounting principles, accounting personnel might object to this allocation. If the costs of poor suppliers are not reflected in reports for the Procurement Department, how will they decide to invest additional effort? It's helpful to include subject matter experts from the production environment who understand the root causes of quality issues.

TURNING COMMITMENT INTO A WORKING PLAN

Once management support is secured, the next step is to develop a structured plan for implementing the quality cost program. Key components of this plan should include the following:

- Present the opportunity: Clearly define the opportunity, demonstrate how the program will generate value, create urgency, and secure leadership's commitment. (More on leadership and strategic use of COQ in Chapter 10.)
- Run a pilot program: Test the quality cost system in a controlled environment to refine processes and validate effectiveness. Beginning to engage a few enthusiastic volunteers to work on breaking down barriers and later help expand the program.
- Train teams: Raise awareness and encourage participation across all functions to build momentum and expand engagement.
- Internal quality cost accounting procedures: Establish standardized methods for tracking and reporting COQ.
- Set up tracking: Ensure accurate measurement of quality costs to drive informed decision-making.
- Integrate with existing quality management systems: Embed quality cost tracking into broader quality improvement initiatives for long-term sustainability.

By following a structured approach, you can transition from initial cost estimates to a fully operational quality cost program that drives measurable improvement. The key is not just to prove the concept but to make quality cost management part of your financial and operational strategy. Use of the quality cost lists in Appendix B will help drive a structured approach.

PILOT FIRST, SCALE WITH CONFIDENCE

For large-scale rollouts, ensuring the long-term success of a quality cost program requires a dedicated, full-time leader. This person should have expertise in quality management, a solid understanding of the company, and a willingness to learn the financial aspects of cost tracking. Ideally, a coleader from accounting should also be involved. The two most important investments in a quality cost program are the leadership team and the quality cost data-collection system—both relatively small commitments compared to the program's potential benefits. These key points should be raised and agreed upon during the management presentation.

After an initial proof of concept, many leaders tend to overestimate the cost and complexity of data collection. However, advances in technology have made this process far more efficient than in the past. For example, one of the pillars of the 4.0 technology revolution is the Internet of Things (IoT). Broadly, this concept means that more and more things will be connected to the internet. For example, a grill that connects to your phone to notify you when the minimum safe meat temperature is reached; a refrigerator that tracks milk usage and reminds users to reorder or reorders milk for delivery itself. In terms of quality, this enables greater connectivity and lower costs across a wide range of sensors. Vibration sensors to monitor preventive maintenance needs and radio-frequency tracking of supplies or tools are just two possible examples. Final inspection and in-process measurements will increasingly be linked to eliminate human time and transcription errors, enabling continuous data recording. In this example, big data, another staple of 4.0 technology, will allow the processing and computation of all this new data. Some systems use artificial intelligence (AI) to support the analysis and reporting of results and the decision-making process for countermeasures. While these are illustrative only, it is not difficult to imagine how sensors and other technologies will track downtime, defects, and other process waste with high precision, thereby enhancing robust COQ processes.

Selecting a Low-Risk, High-Signal Pilot

A pilot program is highly recommended as it serves multiple critical purposes. A pilot program

- demonstrates the system's ability to generate cost-saving results,
- reinforces management buy-in and persuades skeptics of the program's value,
- limits initial implementation scope in terms of time, personnel, and operational areas, and
- allows for system refinement and debugging before full deployment.

Choosing the right pilot area is crucial, as its success will strongly influence the program's expansion. While the scope may range from a single product line to an entire division, the following guidelines can help ensure a high likelihood of success. The pilot area should

- be representative of the company's typical operations,
- align with the organization's strategic priorities or address a high-visibility issue,
- cover all major categories of quality costs, even if some figures are estimated,
- present clear opportunities for cost reduction and process improvement, and
- have cooperative local management willing to support the initiative.

Before diving into detailed planning, an accounting representative should be assigned—either full-time or part-time—to work with the program leader. If a financial coleader has not yet been appointed, this step is essential. This individual will help bridge the gap between existing accounting practices and the quality cost program's needs. Later, they will play a key role in developing the company-wide quality cost accounting framework. The ideal candidate is someone with both financial expertise and an open-minded approach to process improvement. More than just knowing how financial records should be kept, they must understand how they are maintained within the company. Having an accounting representative involved from the outset also strengthens overall buy-in for the program.

There is a difference between common cost accounting practice and the measurement of quality costs. In standard cost accounting practice, all costs are included in reporting. A robust cost accounting system mustn't "cherry-pick" some costs and ignore others. The overriding purpose of a cost accounting system is to ensure all costs are accounted for so that profitability is maintained. For quality cost, excluding some hard-to-measure elements is allowable, as the quality cost reports serve as a guide for internal decisions. This is why the financial coleader must have an open mind.

Building a Team That Can Learn Fast

With a program leader, an accounting representative, and a local management sponsor in place, the core pilot program team is ready. This group brings together all the expertise needed for success. From this point forward, execution requires effort, patience, and perseverance.

Building the pilot program involves several key steps:

- measuring quality costs and establishing appropriate cost bases,
- integrating with fundamental quality performance metrics,
- developing key trend analysis charts,
- identifying and prioritizing improvement opportunities,
- leading problem identification, root cause analysis, and corrective actions,
- ensuring strict follow-through on necessary process improvements, and
- providing clear and concise progress reports.

As the pilot progresses, document it as a case study to support company-wide implementation. Demonstrating how the pilot's results would scale to the broader organization helps leadership visualize the full potential impact. If possible, aligning the pilot program's timeline with the fiscal year end can provide a natural transition point for broader deployment.

CONSTRAINTS, NOT EXCUSES: KEEP COQ MOVING

While full integration with financial reporting systems is ideal, it may not always be feasible. Several potential barriers can limit this integration, including

- financial records structured for compliance (e.g., GAAP) but lacking the internal detail needed for COQ tracking,

- organizational policies restricting access to financial controllers or sensitive cost data, and
- company-specific compartmentalization rules limiting cross-departmental information sharing.

These challenges would not exist in an ideal system, but they may be perceived as roadblocks or real obstacles to overcome. While most challenges can be worked through, when a comprehensive COQ pilot is impractical, practitioners can still add value by implementing a simplified approach.

WHAT YOU MUST HAVE (AND WHAT YOU CAN SKIP)

Even companies with minimal quality management infrastructure often track customer returns and scrap. These two factors alone cover two key COQ categories: waste and downstream consequences. Although a complete COQ analysis is preferable, this basic data can serve as a foundation for initial cost tracking.

The next step should be tailored to the organization's needs, but common areas for expansion include the following:

- Waste—these wastes are often scrap and rework, which may be hidden in operational expenses under production order variances and allowances.
- Monitoring costs—tracking the time and expenses related to inspections, audits (of product), and evaluations of these. Inspections are often a separate account category.
- Investment costs—not to be confused with traditional research-and-development budgets, these investments include quality planning efforts and funding for training and equipment tied to new product development or process improvements.
- Downstream consequences—warranty expenses and customer complaint work are generally monitored in the cost accounting systems, even if only in a simplified form.

With these four categories covered at a basic level, the company will have a more complete view of quality-related costs. While this simplified approach may be less practical in the long run, it can serve as a sensible starting point. Many organizations use this method and continue to extract value from it, proving that COQ tracking—even at a rudimentary level—can drive meaningful improvements. The key is to start simple and continuously refine the process.

EDUCATION THAT ENABLES, NOT OVERWHELMS

Once management approves the quality cost program and the pilot program begins, key representatives from each department should be trained in the principles of quality cost management and the program's implementation plan. This training should emphasize the initiative's cross-functional nature, the importance of teamwork, and the real opportunities for cost reduction and performance improvement across departments.

While most employees have received team training at some point, reinforcing these principles is beneficial, as teamwork is often sacrificed for expediency. Given that the cost of quality management is a new cross-functional effort, collaboration is essential. Concrete examples can help drive this message home. The ultimate goal of this training is to secure buy-in from all departments and a commitment to active participation in the program.

Making COQ Everyone's Work

Each department should have the opportunity to review the entire program and understand its place within the broader framework. As they recognize their expected contributions, they will be better positioned to assess how the program will benefit their specific functions. At this stage, it is essential to encourage department representatives to provide feedback and make recommendations based on their expertise. The quality leaders need to be involved here as advisers, with support from top leadership.

A key exercise is for each department to identify quality cost elements relevant to their operations. This involves listing tasks or functions that would not be necessary if quality were always perfect. Using COQ definitions (see Appendix B), these tasks can then be categorized into investments, monitoring, waste, or downstream consequences.

For example, any activity performed because a product or service does not meet requirements is classified as a waste cost or a downstream consequences cost. If the issue is detected before the product is shipped or the service is delivered, it falls under (internal) waste. If the problem is identified after product release or service completion, it falls into the downstream consequences category. These categorized cost elements become each department's contribution to the quality cost program. Their adoption, along with other departmental insights, helps refine the program's details and ensures each function plays a role in its development.

Not every department will have representative expenses across all four quality cost categories. The responsibility for specific actions may lie outside the scope of a given department's function. Many actions we call investments to improve, monitor, or address downstream consequences have never been considered cost elements under departmental control. This exercise, in which each department identifies its own quality cost elements, may be very illuminating for management. Apply the flow in Figure 3.1 to help sort costs into the four categories.

As more teams engage with the program, additional cost-saving opportunities will emerge. Some suggestions involve revising data-collection or accounting processes, which could become roadblocks to implementation. To avoid unnecessary resistance, focus on the *least arguable value*—a straightforward metric that highlights quality costs without requiring extensive system modifications. The priority should always be fixing the sources of waste, not overcomplicating measurement systems.

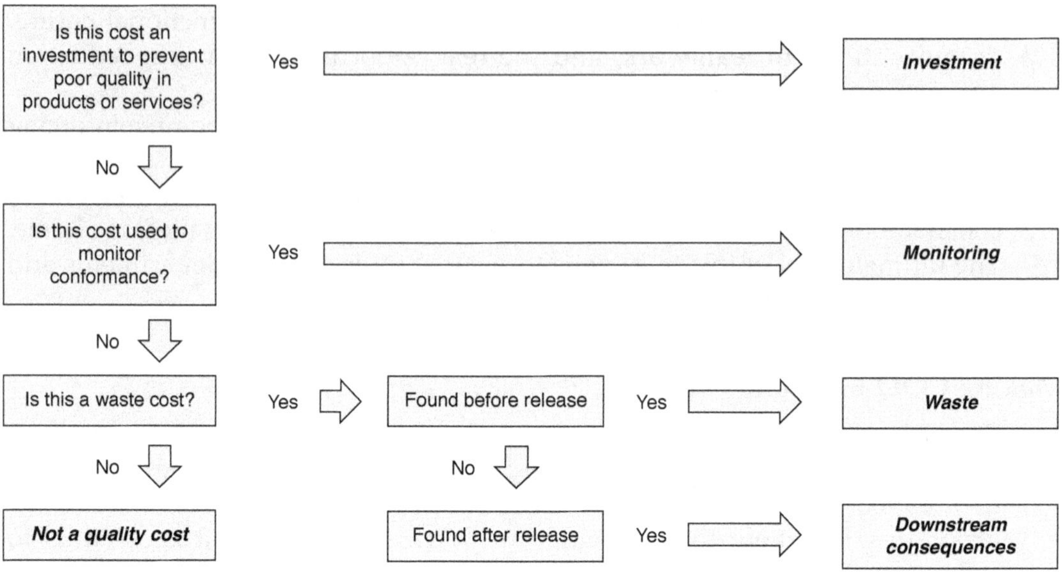

Figure 3.1 Assignment of cost elements to quality cost categories.

Common cost accounting practice is to rely on actual rational cost figures. The idea of using a least arguable value as an estimate may be contrary to how the Cost Accounting Department typically operates. However, there is a difference between broad regular cost accounting reports and COQ metrics. COQ is a process management metric and does not require absolute precision to direct quality efforts where they belong. This makes it essential that the Cost Accounting Department not be responsible for implementing quality cost metrics.

TIE COQ TO WHAT THE BUSINESS CARES ABOUT

Training in quality cost management also serves as an opportunity to reinforce each department's role in overall quality improvement, the benefits they stand to gain, and common challenges to avoid. Key points to emphasize include the following:

- Quality cost management is effective only in organizations with a strong quality system and an improvement program. Without these foundations, COQ tracking serves little purpose. Quality cost efforts should always aim to improve customer satisfaction while reducing costs. While businesses often focus on stakeholder and supplier concerns, a well-rounded COQ approach should also consider employees, end users, and the broader community.
- Corrective action is essential for improvement. Identifying quality costs is only the first step—each department must be fully committed to implementing corrective actions that deliver measurable cost reductions and performance gains.

- Prioritize meaningful cost tracking. Not every minor quality cost element needs to be meticulously recorded. If a cost is not significant enough to warrant monitoring, it is better to omit it rather than complicate the system. The focus should remain on identifying areas where meaningful cost improvements can be achieved.
- Avoid unnecessary complexity in financial tracking. Quality costs should align with existing financial structures rather than forcing unnecessary changes. Adjustments should be made only if the data clearly demonstrates a business need for change.
- Expand the program at a sustainable pace. Even if the pilot program delivers strong results, do not rush full implementation. A phased approach ensures accuracy, consistency, and repeatability—key factors in gaining long-term support and maintaining credibility.

By following these principles, organizations can create a structured, effective quality cost program that engages all departments, drives continuous improvement, and delivers tangible financial benefits.

DOCUMENTING COQ WITHOUT KILLING IT

As the pilot program is formalized, it is an ideal time to establish formal quality cost policies and procedures. The process of writing these procedures ensures logical consistency and helps refine how data is collected and used. One of the early challenges is that much of the necessary quality cost data is not readily available within the existing cost accounting system.

In service industries, for example, many monitoring and waste costs are considered routine operations and are not tracked separately. Similarly, in manufacturing, rework costs may be recorded as variances against a standard allowance rather than as discrete quality costs. These discrepancies, combined with the inherent variation in cost accounting systems, highlight the need for a well-defined internal quality cost procedure tailored to each company's specific financial structure.

Creating a Procedure People Can Follow

A quality cost procedure should clearly define each quality cost element, specify how and when cost data will be collected or estimated, and establish the basis for comparison. To ensure accuracy and relevance, the procedure should also address the inclusion of fringe benefits, overhead allocations, and other necessary accounting adjustments. Additionally, it should assign clear execution responsibilities and provide a standardized format for reporting quality cost data. All document control requirements should be followed, and any applicable RACI (responsible, accountable, consulted, and informed) charts should be updated to reflect roles and responsibilities. (A RACI chart is also known as a *responsibility assignment matrix*. It is a tool that clarifies roles and responsibilities within a project team by categorizing them into four key areas: responsible, accountable, consulted, and informed. It helps prevent confusion and ensures that everyone understands their role in achieving goals.)

For credibility and company-wide acceptance, the procedure should be authorized by the controller or chief accountant. While the quality leader may assist in developing the procedure, its implementation should be a partnership between quality and accounting. If the collection of quality cost data is perceived as a "Quality Department initiative," it risks being viewed too narrowly and not as a company-wide financial priority. In addition to broadening the effort beyond the quality function, the accounting representatives will provide an authoritative stamp of authenticity for the quality cost system.

A Practical Path to a COQ Procedure

A structured approach to procedure development can help integrate quality cost tracking into existing accounting practices:

- Review the company's chart of accounts—identify descriptions of existing cost accounts to determine potential data sources.
- Use established quality cost categories—reference quality cost definitions (e.g., Appendix B) to find formal accounts that align with key cost categories.
- Incorporate department-specific cost elements—utilize input from individual departments (gathered during the quality cost education phase) to identify additional sources of cost data.
- Map quality costs to accounting records—some accounts, like rework, may clearly align with quality costs, while others may require practical judgment. Internal definitions should be simple and meaningful for users rather than overly technical.
- Prioritize consistency over perfection—the primary goal is to reduce all quality costs over time. If a cost falls into multiple categories, prioritize consistent classification over rigid adherence to predefined standards.

For example, if an employee performs multiple tasks—such as a tester who conducts routine inspections (monitoring) but also troubleshoots and retests due to failures (waste)—the company must decide whether to classify their total cost as monitoring or isolate the waste-related costs separately. The most important factor is consistent execution rather than trying to determine the "perfect" category.

When and How to Tighten the Numbers

During procedure development, existing cost accounting practices may be found lacking. You may discover that

- some types of waste (such as scrap) are recorded, while others go untracked;

- customer-related quality costs are high but not clearly and separately accounted for; or
- some quality costs are buried in general expense accounts, making them challenging to analyze.

Addressing these issues can improve financial management and enhance cost visibility. In some cases, refining cost reporting may initially cause quality costs to appear higher, simply because previously hidden costs are now being captured. It is critical to set expectations early so that management does not assume an immediate cost reduction, as monitoring costs are likely to rise as efforts to understand cost drivers begin. Additionally, understanding the typical lag in cause and effect when improving quality will help mitigate unrealistic expectations.

Adapt COQ Definitions to Fit the Business

Each company should tailor its quality cost definitions to its financial structure and improvement priorities. Appendix B provides a detailed description of quality cost elements, which can be adapted as needed. With experience, companies may choose to create, delete, or consolidate quality cost elements.

There is no single "correct" approach to quality cost systems. The best systems are those that

- align with the company's existing cost accounting framework,
- support realistic and actionable quality improvement efforts, and
- evolve to reflect organizational needs and priorities.

By focusing on practicality, clarity, and consistency, companies can develop a robust quality cost procedure that drives meaningful cost reductions and operational improvements.

ASQE Insight on Excellence: Turning Insight into Better Performance

Once quality cost definitions have been tailored to meet business needs, the next step is to operationalize them through data collection. How you collect COQ data will influence not only the accuracy of your reporting but also the credibility of your improvement initiatives. The ASQE/IOE 2025 survey provides a timely snapshot of how organizations are navigating this challenge.

Common COQ Data-Collection Methods

Organizations don't always begin with a fully integrated COQ system. Most start by tracking what's visible and operationally painful. According to the survey,

- 56 percent of companies use *defect document collection,* such as logs that record defect occurrences multiplied by the average cost per defect;

- 48 percent employ *focused assessments,* leveraging surveys or interviews to measure quality costs in specific processes or departments;
- 47 percent rely on *traditional accounting-based methods,* pulling from warranty claims, rework costs, training expenditures, and safety incidents; and
- 44 percent use *time and attendance systems* to capture the hours employees spend on quality-related activities, such as audits or inspections.

The variation is essential. It suggests that there's no universal starting point and that companies tend to follow a predictable path: Begin with tangible events (defects, rework) and expand into systems-based data collection as maturity increases.

Which method fits best with your current systems, and what gaps will it leave?

Internal failure costs are often the first category companies tackle, and with good reason. They're visible, measurable, and often tied to high-impact areas. Survey results confirm this trend:

Internal failure category	% of organizations measuring
Product/service design rework	53
Production or service inefficiencies	52
Supplier/partner rework	51
Retesting or reinspection	49
Scrap and materials waste	48
Supplier/partner replacement	46
Design rework due to deficiencies	42

This focus on rework, scrap, and retesting reflects a pragmatic choice: These costs are often easiest to trace, especially when quality incidents have direct, visible operational impacts. For many organizations, these internal failure costs offer a clear entry point into building a COQ system. They require minimal setup, use data that's often already being collected, and offer fast insight into where processes are breaking down.

From there, a phased approach to growth is the most practical path forward. Most companies begin by tracking basic defect logs and rework reports to establish an initial baseline. These are low-hanging fruit, simple to implement, yet informative enough to guide early improvement efforts. As the system matures, it makes sense to expand its scope. That might mean pulling in data from warranty claims, safety incidents, or training programs, elements that reflect the cost of failure after delivery or the investment made in prevention.

These broader inputs add depth to the analysis and begin to connect operational quality issues to long-term financial outcomes.

Eventually, integrating data systems becomes not just possible but necessary. Connecting quality-tracking tools to time-keeping platforms, Enterprise Resource Planning (ERP) systems, or general ledger accounts enables more accurate reporting, automated updates, and improved cross-departmental coordination. Integration also helps tie operational metrics to financial results, making it easier to justify investments in prevention and continuous improvement.

With a more complete dataset, companies can shift their focus from tracking to targeting. Internal failure data is beneficial for this. If a large portion of quality costs consistently comes from supplier-related rework, that's a clear signal to reinforce supplier qualification processes, renegotiate performance metrics, or increase the frequency of incoming inspections. The data points to where prevention investments will yield the greatest return.

That said, it's important not to let the perfect be the enemy of the useful. Waiting until every data source is clean, consistent, and integrated can delay real progress. Starting with what's available, even if it's incomplete, can still produce meaningful insights. Over time, refinement will come. And it helps to keep finance and operations aligned throughout the process. Finance brings accuracy and validation; operations brings context and relevance. When both sides are involved, the COQ system is more credible and more actionable.

Finally, document your assumptions. If you're estimating defect costs based on averages or proxies, make those calculations transparent. This not only supports benchmarking and audits later but also builds trust in the numbers. Think of your COQ system like scaffolding: It needs to be strong enough to support current initiatives but flexible enough to evolve as the organization grows. The most effective systems strike a balance between analytical rigor and practical usability and, more importantly, are used to inform decisions rather than simply collect data.

FROM COLLECTION TO INSIGHT

Formally preparing and issuing the internal quality cost procedure is essential to ensure control, consistency, and company-wide commitment to the quality cost system. This step is one of the foundational pillars of a mature, comprehensive quality management system.

Capturing COQ Without Creating Busywork

Ideally, the internal procedure should include a structured system of quality cost elements that enables easy classification and reporting of investment, monitoring, waste, and downstream consequence costs. These cost elements should be coded, so they can be sorted and aggregated clearly. The coding system outlined in Appendix B illustrates this structure.

A well-designed coding system allows for quick calculations, such as follows:

- All cost elements coded "1" represent the total investment for prevention costs.
- All cost elements coded "2" represent the total monitoring cost.
- Additional digits can be used for more granular breakdowns (e.g., "1.3" for investment for prevention costs related to purchasing).

Using this structured coding approach makes it more efficient to collect labor-related quality costs. Employees can enter applicable quality cost element codes on labor distribution forms, time cards, or digital entry systems. These entries, in turn, are processed by the company's operational software.

One exception to real-time data collection is scrap reporting. Since scrap is often not identified at the time of production, it must first be inspected, rejected, and dispositioned before being recorded. In many companies, scrap-reporting documents are forwarded to estimating teams, who assess the labor and material costs incurred up to the point of scrapping. This differs from replacement cost, which is the total cost to complete the job. In quality cost tracking, the focus is on the actual cost of loss up to the point of scrap rather than the total cost of redoing the work.

Finding Signal in the Numbers

Once a system for collecting quality costs is in place, the next step is to organize and analyze the data. Quality management software systems may seem ideal to, and be marketed to, efficiently gather costs, such as scrap—or Corrective and Preventive Action (CAPA)—processing time. Still, at this time, many popular systems do not capture quality costs in any distinct and meaningful way. While some are nearly limited to the example given, none of the digital systems investigated (investigation conducted at the ASQ World Conference on Quality and Innovation 2024) were found to advance COQ data collection or analysis robustly. Despite digital advancements in the 4.0 era, a spreadsheet-based approach remains the most practical starting point, as it offers flexibility and easy error detection without requiring complex programming. The process typically involves the following:

- Compiling raw data—quality cost elements are listed in a spreadsheet alongside the departments, areas, or projects where the costs occur (Figure 3.2). This is used by accounting each reporting period to track which costs are being recorded—and which may be missing. This record may help identify when and where extreme values occur due to a data system issue.
- Summarizing collected data—the reported costs are collated into a summary sheet (or a dedicated tab within a workbook) to present the information in a format aligned with the intended use (Figure 3.3). This summary serves as the primary quality cost report used by the Quality Department.

- Identifying trends and improvement opportunities—over time, quality cost data should be analyzed alongside basic quality performance metrics to identify patterns, trends, and areas for improvement.

Element (V) / Department (>)	Accounting	Administration	Engineering	Estimating	Marketing	Production	Quality	Shipping
Code and description								
1.1.1 Marketing research								
1.1.2 Customer/user surveys/clinics								
1.1.3 Contract/document review								
1.2.1 Design quality progress reviews								
1.2.2 Design support activities								
1.2.3 Product design qualification test								
1.2.4 Service design qualification								
1.2.5 Field trials								
1.3.1 Supplier reviews								
1.3.2 Supplier rating								

Figure 3.2 Quality cost data spreadsheet.

While spreadsheets offer an easy-to-modify, transparent way to manage quality cost data, they require ongoing human oversight. If the system proves stable, transitioning to an automated software solution may improve efficiency. However, in highly dynamic environments where priorities frequently shift, a spreadsheet may remain the best option due to its adaptability.

From Insight to Improvement

Initially, quality cost data should be analyzed over a sufficient period to establish a baseline, verify current improvement opportunities, and provide

Description	Current month			Year to date		
	Quality costs	As a % of Sales	As a % of Other	Quality costs	As a % of Sales	As a % of Other
1.0 Investments						
1.1 Marketing/customer/user						
1.2 Product/service/design development						
1.3 Purchasing prevention investments						
1.4 Operations prevention investments						
1.5 Quality administration						
1.6 Other investments						
Total investments						
Investment targets						
2.0 Monitoring costs						
2.1 Purchasing supplier monitoring costs						
2.2 Operations monitoring costs						
2.3 External monitoring costs						
2.4 Review of test and inspection data						
2.5 Misc. quality evaluations						
Total monitoring costs						
Monitoring targets						

Base data	Current month Budget	Current month Actual		Year to date Budget	Year to date Actual	
Net sales						
Other base (specify)_______						

Figure 3.3 Quality cost summary report.

actionable insights for teams. These insights should be shared with work groups and leadership to ensure collective understanding and commitment to corrective actions. Once problem areas are identified, specific improvement targets should be set.

Since quality cost tracking is a tool for quality and process management, it should integrate seamlessly into broader quality improvement efforts. This data is typically used to

- develop trend charts that illustrate initial quality cost baselines, improvement targets, and actual progress (Figure 3.4) and
- generate monthly or quarterly progress reports that align with overall quality management goals (Figure 3.5).

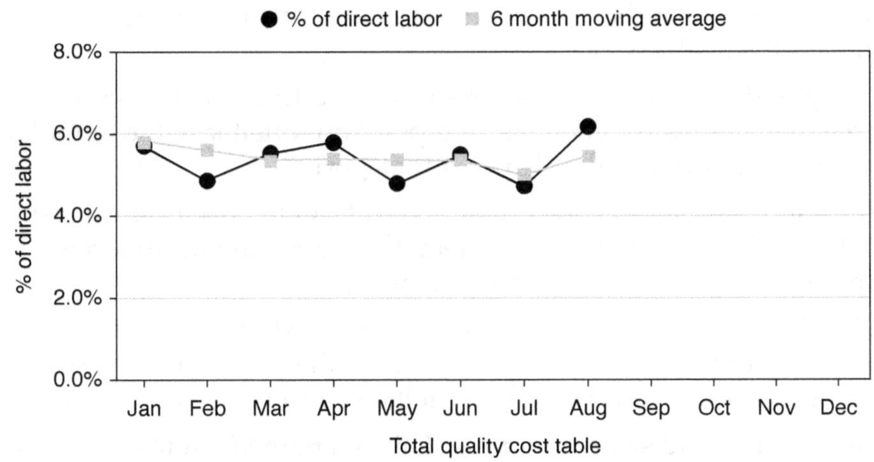

Figure 3.4 Total quality costs.

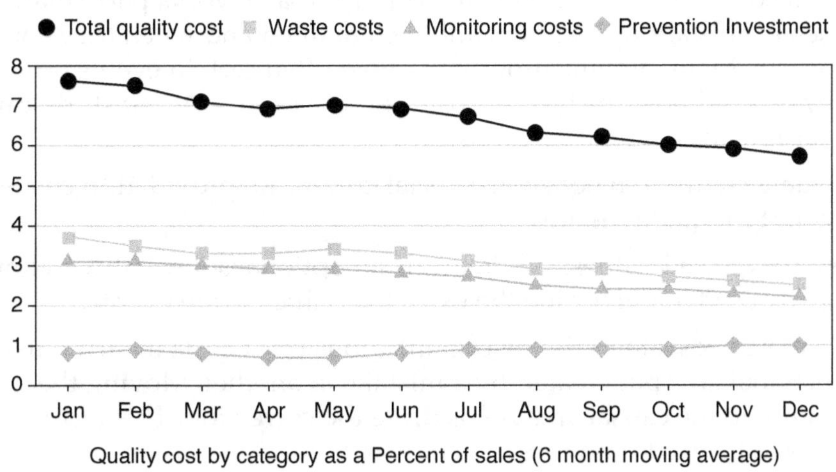

Figure 3.5 Costs related to quality.

More on using COQ for improvement is found in Chapter 5.

MORE THAN ONE WAY TO CAPTURE COQ

There are multiple ways to collect quality cost information. Some of these come from cost accounting systems, others from existing process measurements, and some may need to be created to provide detail for the COQ system. What's most important to remember is that, because of differences across organizations, the COQ measurement is always customized. Category totals may be benchmarked across organizations, but the details within each will always vary.

- Using budget and expense sources: Sourced from existing cost accounting accounts, these are a good starting point but often lack the granular detail needed to illuminate the four COQ categories.
- Additions and subtractions from accounts: Often used due to insufficient detail in the cost accounting system, these adjustments to reported amounts clarify the four COQ categories.
- Yield losses from existing process cost reports: Most organizations have some form of process waste measurement. While this information is detailed, there may be a tendency to overlook the dollar value of these yield losses.
- Head counts and salary amounts: This is a refinement to the additions and subtractions listed above. Particularly useful for measuring investment in improvement, care must be taken not to list individual salaries explicitly. A better practice is to use estimates of typical salary levels for job titles.
- Event counts and cost per event: This approach works particularly well when events are well known and well defined and when they contribute to waste. Many organizations track how often certain events occur, but they do not know the typical costs of these events. A cost study can determine those typical expenses.
- Notable credits: On occasion, special credits are provided to customers to account for quality issues.
- Using expected values: While this is an optional source, this technique may be used to determine the expected value of rare events.
- Charge-back systems: Some organizations track costs incurred in one functional area but created by a mistake in another. Moving the event expense to the causal area helps drive the correction. This technique needs to be used with care.
- Manual corrections: No system can measure costs perfectly. When significant errors occur, it helps to include manual corrections driven by human oversight. These corrections need to be tracked as part of the COQ procedure.
- Comparison of best case to actual cost: Sometimes referred to as the *Taguchi method,* this technique involves significant sophistication to

determine the magnitude of best-case costs.

COMMON TRAPS (AND HOW TO AVOID THEM)

One of the most significant risks in implementing a quality cost program is treating it as a standalone initiative rather than integrating it with the broader business strategy. Quality cost reports—no matter how detailed—are meaningless without a clear connection to actual company performance.

To ensure meaningful results, quality cost tracking should be directly tied to tangible business outcomes, fostering an ongoing dialogue among quality teams, operations, and leadership. By embedding COQ analysis into decision-making processes, companies can move beyond measurement and into meaningful, data-driven improvement. Frequently, a scrap cost is found downstream from the causal area. A team approach can help reduce downstream costs through upstream investments.

A second pitfall is treating lost sales revenue as a quality cost. Besides the fact that revenue includes margins and thus compares poorly with operational costs, if sales staff begin blaming operations' poor quality for lost sales, it creates an unfortunate internal conflict. Sales and operations need to communicate effectively and often. How is a lost sale to be tied to a specific mistake or failure? Customers will rarely drop a supplier for a single instance of poor quality; if the supplier cannot respond to the customer comprehensively and promptly, is the lost revenue due to the inability to respond or to the initiating failure? The full consequences of blaming need to be considered. The goal of any metric is to drive the correct behavior. What behavior is being promoted?

ACTIVITY-BASED COSTING: HELPFUL TOOL, NOT A REQUIREMENT

Activity-based costing (ABC) is a topic used because it addresses a real need. Traditional accounting does not naturally show how costs break down within real manufacturing systems, which limits visibility for business decision-makers. ABC improves visibility by linking costs to specific activities, such as setup time, machine usage, and labor. As a result, many organizations use ABC in everyday accounting and business decisions. For COQ, the visibility gaps can prompt quality and finance teams to collaborate to interpret costs buried in standard accounts. From a quality perspective, ABC makes inspection, rework, and prevention costs easier to identify and discuss.

The practical takeaway is to understand what ABC offers and apply it where it improves decision-making. Additional detail and supporting examples are included in Appendix A for readers seeking further context.

COQ IN A DIGITAL, CONNECTED WORLD: SOFTWARE AND CYBERSECURITY

Industry 4.0: A New Operating Context for COQ

COQ has long been a focus of academic research and organizational strategy, as companies strive to balance investment, monitoring, and failure costs—both waste and downstream consequences. Now, with the rise of industry 4.0, manufacturing 4.0, and service 4.0, the way these costs are measured, managed, and optimized is evolving faster than ever. The integration of cyberphysical systems into production and service processes presents both new opportunities to reduce COQ and new challenges in accurately tracking and allocating costs.

From Manual to Integrated COQ

Key industry 4.0 technologies—such as IoT, cloud computing, AI, machine learning, digital twins, big data, and 3D printing—are transforming how organizations identify and control quality costs. Some technologies, such as cloud computing, primarily affect cost structures by reducing data-storage and processing expenses. Others, such as IoT and digital twins, enable real-time process monitoring, significantly improving failure detection and prevention.

In service industries, AI-driven systems are revolutionizing defect detection, customer complaint analysis, and service failure tracking, enabling more accurate measurement of downstream consequence costs.

Service industries also incur office waste or quality-related office costs. AI significantly impacts office environments by dramatically accelerating knowledge acquisition. General-purpose large language models may add efficiencies. As more organizations adopt customized, specialized AI that leverages their own processes and documentation, it becomes easier to identify areas of conflict and confusion. One of the significant sources of waste in office work is poor workflow. By identifying and codifying workflows and streamlining processes, AI can aid in substantially reducing office waste. The investment necessary to implement a custom large language model is a cost of good quality. It's also essential to identify additional monitoring activities to ensure that these models do not hallucinate solutions. See more on the cost of office waste in Chapter 8.

Whether in manufacturing or the office, these tools are already used across industries, and while their impact on COQ may not seem direct, it is growing. Examination of principles may be highly effective. One thing is sure: Technological acceleration will continue, making proactive adaptation essential.

Key Actions: What to Do Differently in a Digital Environment

To take full advantage of these advancements and effectively manage quality costs, organizations should focus on four key areas.

1. Strengthen Cross-Disciplinary Collaboration

COQ management is no longer just a function of the Quality Department. It now requires close coordination with IT, finance, and operations to integrate industry 4.0 technologies into cost-tracking and reduction strategies. Quality professionals must work with ERP administrators and data scientists to ensure COQ data is accurately captured, analyzed, and leveraged for decision-making.

2. *Implement Data-Driven Decision-Making in COQ*

Quality professionals are no strangers to data. Industry 4.0 technologies will generate vast amounts of quality-related data, but without structured analysis, this data will stay underutilized. To optimize COQ measurement and control, organizations should

- use real-time data monitoring to reduce monitoring costs and internal waste,
- apply predictive analytics to anticipate defects and lower all quality costs, and
- leverage AI-driven quality audits to improve the accuracy of COQ reporting.

3. *Upskill the Workforce for Digital COQ Management*

A workforce unfamiliar with AI, automation, and data analytics may struggle to leverage industry 4.0 to reduce quality costs fully. To bridge this gap, organizations should

- offer training on emerging technologies that impact COQ measurement and control,
- foster a culture of continuous learning to ensure employees can adapt to evolving quality cost management techniques, and
- hire data-literate professionals to enhance COQ analytics capabilities.

4. *Align COQ Strategy with Business Transformation*

COQ management should be integrated into digital transformation initiatives to ensure that industry 4.0 investments directly contribute to quality cost reduction. While this is explored further in Chapter 10, executives should consider

- embedding COQ metrics into company-wide performance tracking,
- ensuring suppliers and partners align with digital quality cost management, and
- developing adaptive COQ frameworks that evolve alongside technological advancements.

Looking Ahead

The way organizations calculate, track, and optimize COQ is changing. Industry 4.0 is no longer a theoretical concept—it is embedded in every aspect of modern organizations. As a result, quality management strategies must evolve to incorporate digital tools and analytics. Companies that successfully integrate these technologies will not only improve quality but also lower failure costs, enhance prevention efforts, and strengthen their bottom line.

Managers who master these changes—and the broader practice of adaptation—will be best prepared to support themselves and their organizations in the rapidly shifting, sometimes unpredictable realities of the future.

RISK: PROTECTING THE COQ PROGRAM AND BOTTOM LINE

Typically, quality initiatives are most challenged by resource constraints and opportunity costs rather than by tangible downside risks; however, as risk-based thinking gains broad acceptance, it warrants analysis. Implementing a COQ program introduces risks that can impact accuracy, adoption, and long-term effectiveness. Some risks, such as data inconsistencies or employee resistance, are easy to anticipate, while others—like supply chain dependencies or cybersecurity threats—may be less obvious. Addressing these risks early ensures that COQ efforts lead to meaningful improvements rather than becoming a burdensome administrative task.

Risks That Undermine COQ

The first set of risks relates to the internal execution of the COQ program. Even when an organization understands the value of tracking quality costs, implementation challenges can lead to gaps in data collection, inconsistent reporting, and low adoption rates. Common risks include the following:

- Inaccurate data collection: If different departments classify quality costs inconsistently, COQ metrics will be unreliable. Training and process sharing are helpful here.
- Poor integration with existing processes: If COQ tracking is treated as an additional administrative task rather than embedded into daily operations, compliance will suffer. Look for ways to simplify data handling.
- Lack of executive buy-in: Without visible leadership support, COQ programs often fail due to insufficient resources or insufficient prioritization. Be sure the successes are communicated.

On a broader scale, COQ tracking can also expose organizations to financial and regulatory risks:

- Misalignment with business goals: If COQ reporting is not integrated into strategic decision-making, it may become an isolated reporting exercise rather than a driver of improvement (see Chapter 10 for more on aligning COQ with business objectives).
- Financial and regulatory exposure: Errors in COQ reporting can impact financial forecasting, compliance audits, and operational decisions. Rapid identification and correction of mistakes is key, with proper explanations of how the errors occurred and how they were corrected.

Why People Push Back: Change, the Overlooked Risk

One of the most underestimated risks in COQ implementation is organizational resistance to change. Employees accustomed to traditional quality tracking methods may be skeptical of new reporting structures, question the accuracy of data-driven assessments, or quietly fear increased accountability.

To mitigate resistance, organizations should take a structured approach to change management. Frameworks such as Kotter's 8-Step Change Model and Lewin's Force Field Analysis provide proven strategies for guiding teams through transitions. Many of the steps suggested in this chapter generally align with Kotter's steps. Still, the specific model is less important than the overall approach; leadership must ensure employees understand why the changes are happening and how they benefit both the organization and individual teams.

For example, if frontline workers fear that COQ tracking will be used to penalize them for defects, leadership must emphasize that the focus is on process improvement, not individual blame. Clear communication, employee training, and active involvement in the process will reduce resistance and increase engagement.

Supply Chain and Data Risks

Supply chain dependencies can introduce hidden quality costs that are difficult to capture in standard COQ tracking. A poorly managed supply chain can lead to the following:

- Underreported failure costs: If suppliers fail to provide detailed defect and rework data, hidden quality costs go untracked.
- Inconsistent quality standards: Global suppliers may have variable quality controls, leading to unexpected warranty and failure costs.
- Third-party software dependencies: External platforms used to collect or analyze COQ data may not align with internal methodologies, creating errors.

See Chapter 6 on suppliers' COQ for more information.

Mitigating Supply Chain Risks

To reduce these risks, organizations should

- communicate honestly with the supply chain with an eye for mutual benefit,
- mandate standardized quality cost reporting from vendors,
- include quality cost clauses in supplier contracts, and
- evaluate third-party software providers to ensure alignment with quality management and cybersecurity requirements.

Cybersecurity: Cyber Risk Is Business Risk

As COQ data increasingly moves to digital platforms, cybersecurity must be a priority. Unauthorized access, data manipulation, or breaches can compromise the integrity of COQ reporting, leading to

- unauthorized access to sensitive information,
- quality financial misstatements, and
- operational disruptions to corrective processes.

Protecting Integrity and Access

Organizations should

- implement strict access controls to prevent unauthorized data modifications,
- use encryption and secure cloud storage to protect quality cost records, and
- regularly audit cybersecurity policies to ensure COQ systems remain protected from external threats.

Capability Maturity Model Integration and COQ: Security as Capability, Not Afterthought

The Capability Maturity Model Integration (CMMI) framework provides a helpful structure for evaluating an organization's maturity in managing performance. Originally around software development, the principles can be applied to deploying COQ securely. Applying CMMI principles ensures that

- COQ tracking is systematically embedded into business processes rather than being reactive,
- data governance policies are robust, preventing security lapses, and
- cybersecurity practices scale with the organization's digital reliance on quality tracking.

CMMI principles can also be applied to supplier security assessments, ensuring that vendors handling quality data meet cybersecurity standards. Organizations that integrate cybersecurity into their COQ strategy will be better positioned to protect data, maintain compliance, and ensure the long-term integrity of their quality cost tracking systems.

Risk Awareness Without Paralysis

Risk is inherent to COQ implementation, but it can be managed through proactive planning, robust data governance, and structured change management. Organizations that address these risks head-on will avoid costly disruptions, improve the accuracy of COQ reporting, and enhance their overall quality cost management strategy.

CONCLUSION: PUTTING COQ TO WORK

Implementing a COQ system is not a one-time initiative—it is an ongoing effort that must evolve alongside an organization's quality management practices. This chapter has explored the fundamentals, risks, and challenges of implementation, emphasizing the importance of structured planning, organizational buy-in, and cross-functional collaboration.

The first major challenge is simply getting started. As discussed earlier, developing a COQ program requires both financial and operational understanding, as well as the ability to secure leadership support. The early stages—gathering initial quality cost estimates, identifying major cost drivers, and demonstrating the financial impact—create the foundation for a practical, structured approach to quality cost management. Once management is on board, the focus shifts to building internal processes, piloting the system, and embedding COQ into everyday decision-making.

From there, the work does not stop. A COQ system must be maintained, refined, and actively used to drive continuous improvement. As detailed in the following chapters, simply tracking costs is not enough— the insights gained must inform better business decisions, whether by reducing failure costs, improving prevention efforts, or aligning quality investments with company strategy. COQ is not just a financial tool; it is a strategic asset for enhancing the quality, efficiency, and profitability. Chapter 10 dives deeper into the relationship between COQ and strategic management.

The journey from COQ implementation to continuous improvement mirrors the natural progression of quality management maturity. While the initial implementation phase establishes the foundation, the actual value of COQ comes from ongoing application, refinement, and integration into broader business strategy. By adopting a structured, adaptive approach, organizations can transform COQ tracking from a routine administrative task into a powerful driver of financial and operational success.

TOOL: THE LEADER FRAMEWORK FOR SCALING A QUALITY COST PROGRAM

- *Lead with a champion and an accounting colead.*
- *Estimate the opportunity using the least arguable values.*
- *Align leadership on what "credible" means and what decisions COQ must support.*
- *Design a low-risk, high-signal pilot with clear selection criteria.*
- *Embed procedures, definitions, and responsibilities, so they survive scale.*
- *Run the cadence, trends, reporting, and phased expansion.*

TOOL: STAKEHOLDER ANALYSIS GUIDE

Identify your key stakeholders and analyze their interests:

Who has information, expertise, or experience that could meaningfully support this project?

Whose support is needed to improve the chances of a successful outcome?

Who could inhibit our success?

Who serves as our champion in the organization?

Who can help communicate how the project aligns with the enterprise's broader goals?

Who is paying for the project?

Who will work with us to implement the project, from our team, other teams, or organizations?

Who will benefit from the results?

Who needs to know about our COQ outcomes that will impact their work?

Who will have to change the most?

Categorize how involved each key stakeholder needs to be:

Are they

1. High Influence / High Perceived Value? Then manage them closely. Fully engage them and ensure you meet their expectations.
2. High Influence / Low Perceived Value? Then seek to keep them satisfied. Put enough effort into these people to keep them satisfied but not so much that they become disinterested in the COQ project.
3. Low Influence / High Perceived Value? Just keep them informed. Talk to them periodically to ensure that they have no significant issues. (They can often be invaluable with the details and implementation.)
4. Low Influence / Low Perceived Value? Monitor these stakeholders and avoid burdening them with excessive communication.

Map positions / current support of key stakeholders:

For each group above, make a chart on where they stand now and where they need to be for the COQ project to succeed. Do not show them this chart! An O is where they are now; an X is where they need to be.

Stakeholder	Unaware	Aware	Understands	Supports	Committed
A	O			X	
B		O	X		
C					O, X

Make a plan to influence the right people in the proper manner:

Using the map above, list for each stakeholder who needs to be influenced:

- Who influences whom
- Individual level of resistance (high, medium, low)

- Their issues, concerns, interests
- Key messages to communicate
- How to communicate
- When to communicate
- Who does the communication

KEY TAKEAWAYS

- *Use the LEADER Framework* to guide any COQ implementation.
- *COQ needs a champion with credibility.* Someone has to own the push, understand the system, and speak finance well enough to earn trust in the room.
- *Do the homework before you ask for money.* Start with what is already visible (scrap, rework, returns, warranty), add reasonable estimates, and then point to the most significant improvement targets.
- *Win leadership by involving them early.* Ask what they need to see to say yes, then design a focused follow-up study that proves the complete cost-reduction cycle.
- *Build a cross-functional team, not a "quality project."* Partner with accounting and operations and be willing to report costs to the process owner, even if that differs from the classic accounting structure.
- *Pilot first, then scale with a plan.* Choose a low-risk, high-signal pilot, document it as a case study, and expand at a pace that protects credibility.
- *Keep measurement practical.* Use the least arguable value when perfect data is not available, track only what matters, and tie COQ to corrective action so reporting does not become busywork.
- *Modern COQ includes digital risk.* Industry 4.0 tools can improve detection and reduce failure costs, but cybersecurity and data governance become vital if you want leaders to trust the numbers.

Chapter 4
Quality Control with COQ

Ever wondered why your hard-won improvements fade after several months? The answer often lies in control, and quality cost data is your best ally. This chapter shows how to use COQ not just for improvement but for sustaining gains through effective control.

Juran's *quality trilogy* highlights three essential components: *planning, control,* and *improvement*. The previous chapter focused on planning—specifically, on establishing a COQ system as the structural foundation for tracking, managing, and reducing quality costs. After a system is in place, organizations naturally turn to improvement. But too often, they skip over control—the quieter, behind-the-scenes discipline that ensures gains are maintained.

Control is rarely ignored entirely, but like cost avoidance, it is often undervalued. Without it, even the most successful corrective actions risk being lost to drift or regression. "Why spend on control if there are no pressing issues?" is a short-sighted viewpoint.

CONTROL: HOW YOU KEEP THE GAINS

In Juran's model, stabilization precedes and follows every change. Control is what secures the gains of improvement. A COQ system plays a key role in this function—it monitors for drift, provides early warning signals, and supports data-driven decisions when performance strays off course.

Steps to Implement Control Using COQ

1. *Choose the control subject*
   Select the target category, department, or process.
2. *Define measurement methods*
   Define and document the measure, source, and update cadence.
3. *Set performance standards*
   Similar to control limits, set a baseline range and a drift trigger.
4. *Measure actual performance*
   Review the trend routinely according to the established method.

> 5. *Act on deviation*
> Execute the reaction plan primarily by acting on special cause deviations.

Think of it this way: improvement is acceleration; control is steering and braking. Without control, speed becomes a risk.

In practice, control enables organizations to establish and maintain a baseline. Without knowing what "normal" looks like before intervention, it's impossible to measure the success—or stability—of an improvement. COQ data helps establish the baseline and confirms whether changes deliver lasting results.

FIX IT ONCE: CORRECTIVE ACTION THAT PAYS BACK

Control is not just about avoiding chaos—it enables more intelligent decisions. When a COQ system identifies a costly issue, it supports targeted corrective actions. These actions must be justified, of course. A $5,000 fix for a $200-per-week rework issue may seem excessive—until it becomes clear that doing nothing is costing $10,000 annually.

Some problems are easily resolved: updating a tool, tightening a spec, clarifying a work instruction. Others are harder to diagnose: design flaws, vague responsibilities, or marginal process conditions. Even harder are issues that result from multiple causes that initially seem unrelated. Without structured, ongoing cost metrics, these issues are often dismissed as the "cost of doing business."

For example, a manufacturer observes rising customer complaints and identifies a common thread in packaging failures across shipments. At first glance, there are many incorrect labels on shipments. A sweep is made of customer databases and other label information datasets. Many mistakes are found and fixed. After six months, other problems come to light: cheaper packaging materials that failed to withstand shipping in subzero temperatures, incorrect power cords on shipments to distant countries, and instructions too small for customers to read that fail to convey critical startup information. As these issues are found, timing-adjusted COQ data show that three retirements of essential employees were responsible for all of them. The firm realized it had no effective knowledge management system, and key knowledge was being lost.

A functioning control process brings these hidden issues to the surface early, when they're still manageable and when the cost of action is still lower than the cost of delay. The principle remains timeless: You pay once for corrective action, but you pay indefinitely for inaction.

Shared Ownership Is What Makes Control Work

It's easy to assume that quality failures begin on the shop floor. In reality, many originate from other areas, such as engineering, planning, tooling, and documentation. Even well-intentioned upstream decisions can create costly downstream issues. Quality costs in office environments may differ from those in production areas, but errors in the office can lead to expensive problems in production. Control extends visibility across functions. A well-structured COQ system helps shift accountability from individuals to systems, replacing blame with cross-functional problem-solving.

COQ should not be viewed as a tool for quality professionals alone. It belongs in the broader management conversation, involving finance, operations, engineering, and even human resources (HR), where training and procedural adherence are key.

As Dale and Plunkett (1995) and later Schiffauerova and Thomson (2006) noted, quality cost includes "the costs incurred in the design, implementation, operation, and maintenance of a quality management system[,] . . . the cost of resources committed to continuous improvement, the costs of system, product, and service failures, and all other necessary costs and non-value-added activities required to achieve a quality product or service." In this context, control is not merely about preventing regression—it's about coordinated action across the entire value stream. Managing a value stream requires more than budgetary adherence when cross-functional issues arise. It's not an academic exercise; real process management cost data, including investments in monitoring and waste, are required.

Making Quality Costs Visible to Management

Control also plays a critical role in management reporting. For managers, COQ metrics aren't just reports—they're decision tools. They show where to invest, where to hold steady, and where to intervene fast. They allow leadership to identify persistent issues and evaluate which areas are improving and which are not. We are talking about data for decision-making and for current and future process control.

But these metrics must be interpreted wisely. Not all quality costs are comparable. For example, a new, complex product line will naturally carry higher costs than a mature, optimized process. The more useful perspective is directional:

- Are costs decreasing over time?
- Are improvements holding?
- Are we maintaining performance after interventions?
- Is additional monitoring required to determine performance trends?
- What are the correlations between investments for quality and waste / downstream consequences?

When used correctly, COQ reporting keeps the organization focused on sustainability, control, and long-term progress.

TRENDS TELL THE STORY

There are two types of quality cost trend analyses: long-range and short-range. Long-range analysis typically views total quality costs over quarters or years. It is used primarily for strategic planning and monitoring overall progress.

Short-range trend charts are prepared for each company area, where individual quality cost improvement goals are to be established. One approach is to assign a short-range target to each general operational area simply. Alternatively, an initiative can become as detailed and sophisticated as the quality management team and system will support.

Figure 4.1 is a composite example of a long-range quality cost trend chart for a typical organization with sales of $100 million to $200 million. It shows the total COQ as a percentage of net sales over ten years. It also separately reports investment, monitoring, waste, and waste plus downstream consequence costs as a percent of sales (downstream consequence costs are shown as the difference between waste and waste plus downstream consequence costs). The first two years show quality cost history without any knowledge of or emphasis on its reduction. The third year is the start of quality cost measurement and use. Years 4 through 9 show actual progress accomplished. Year 10 is a projection of the expected continued progress. Note that a short-term drop in sales occurred in year 4, effectively hiding improved quality cost performance as a percent of sales.

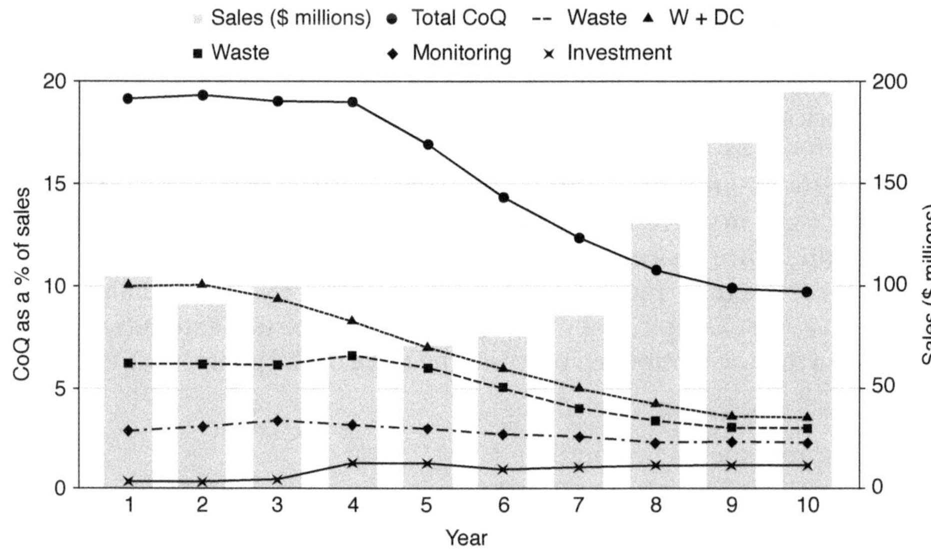

Figure 4.1 Cost of quality trend analysis.

A minimum-quality measurement system should include summary monitoring results for all key operational areas. These will typically be receiving inspection, fabrication inspection, final assembly inspection and test, and field monitoring reports. The summaries are usually presented as trend charts to indicate and make people aware of current quality performance levels. Quality cost trend charts, when correlated, will supplement these performance charts with viable cost data to support the improvement effort. This is the essence of their use together. (See Figure 4.2 for a rework cost example.)

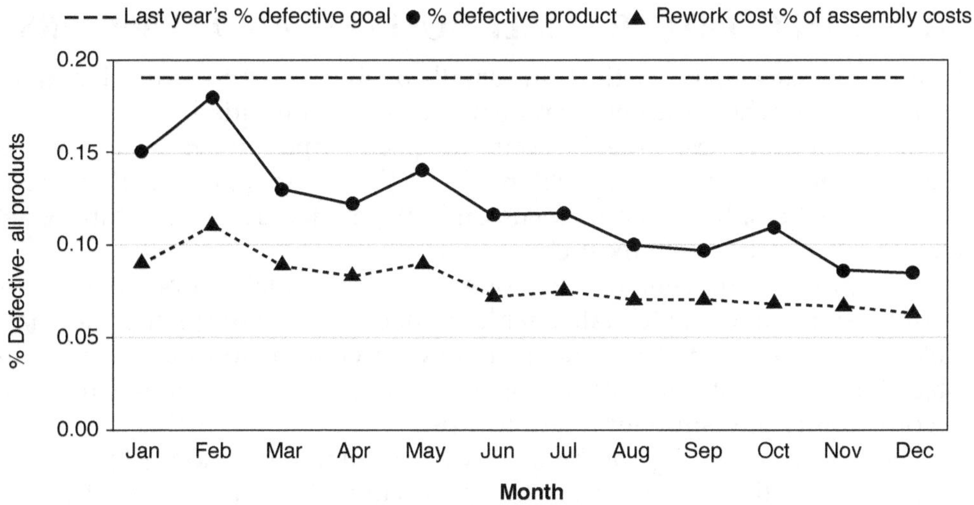

Figure 4.2 Assembly area quality performance.

There is usually a time lag between basic quality measurement data and quality cost data. Quality measurement data is always current (typically daily), whereas quality cost data is accumulated retrospectively in cost accounting reports. It is crucial, therefore, to understand that quality costs are used to support before-the-fact improvement and to verify its after-the-fact accomplishment. Still, actual improvement stems from using current quality measurement data to identify root causes and implement corrective actions.

There is also a time lag between cause and effect. (Quality improvements do not show immediate reductions in quality cost because of the time lag between the cause and its impact.) This lag is evident in a quality cost trend chart. For this purpose, it may be desirable to indicate on the chart when quality improvements were made. Figure 4.3 is a simplified quality cost trend chart marked to indicate the start of a quality improvement activity. The note explains the steady progress over the last five months: In April, training programs were initiated.

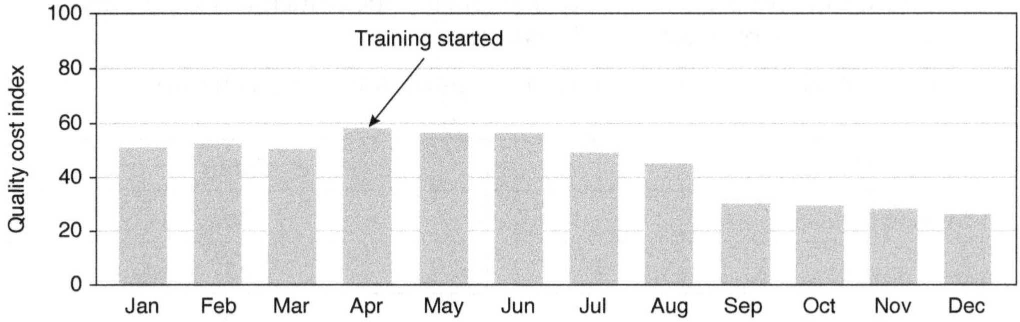

Figure 4.3 Quality cost trend.

CONTROL IN CONTEXT: SMALL BUSINESS AND SUPPLIERS

Control is especially critical in small businesses and supplier quality management, where the cost of repeated errors is amplified.

In small businesses, limited resources make it imperative to avoid low-value fixes and prevent recurring problems. A COQ system, paired with consistent control practices, helps teams focus on high-impact areas—avoiding wasted effort and ensuring each fix sticks.

In supplier management, control ensures that quality does not begin at your receiving dock. COQ data enables you to measure supplier performance, detect early signs of drift, and pursue corrective action based on shared, objective data—not assumptions or blame. It builds trust, transparency, and accountability into supplier relationships.

There is no better way for the quality management function to "put its stake in the ground" than to develop a strategic quality plan substantiated by quality costs and to have this plan committed to in the organization's business plan.

Similarly, cost control from a supplier quality perspective is explored in the subsequent chapter on supplier quality, where it can be more effectively integrated with improvement strategies and addressed in a more comprehensive, focused way.

PROBLEM-SOLVING TOOLS THAT TURN COQ INTO ACTION

We have mentioned several tools that may be used to provide meaningful interpretations of COQ. It may be helpful here to review those tools and understand how their use significantly affects an organization's benefits.

Pareto Analysis: Finding the Vital Few Costs

The Pareto Principle is named after Vilfredo Pareto, an Italian economist and sociologist from the late nineteenth century. Looking at land ownership, he found that 20 percent of individuals owned 80 percent of the land. He made a second discovery in his garden: 20 percent of the plants produced 80 percent of the fruit. And while the Pareto Principle is named after him, he did not develop the Pareto chart as we know it today. The chart was created and popularized by Joseph M. Juran in the 1940s, as he recognized the principle's generality.

Let's build a simple chart to illustrate:

1. A table of data is accumulated, perhaps of defects by category.

Category	Count
Missing parts	29
Out of spec	25
Damaged parts	10
Customer damage	5

2. These categories are arranged in descending order in the vertical bar chart in Figure 4.4, with their count frequencies on the left.
3. To make the final Pareto chart in Figure 4.5, create a cumulated percentage of the counts and plot a line over the bar chart with a secondary axis on the right side.

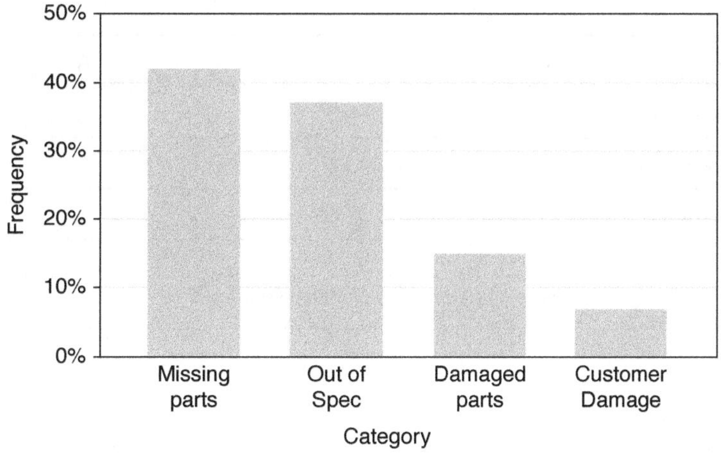

Figure 4.4 Incipient Pareto chart.

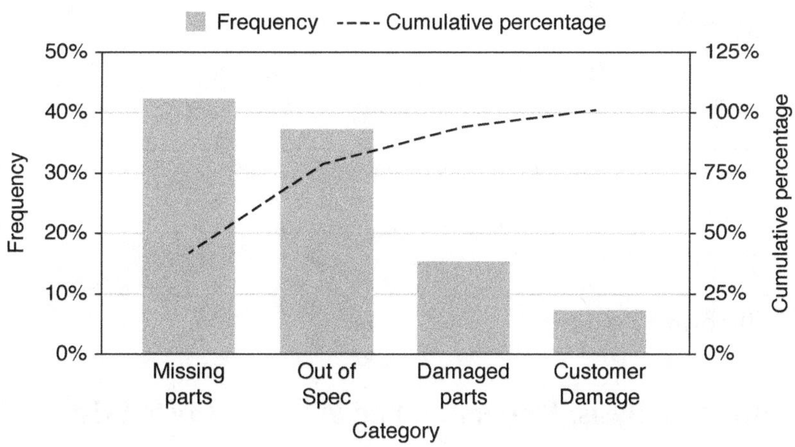

Figure 4.5 Final top-level Pareto chart.

In Figure 4.5, the first two bars ("Missing Parts" and "Out of Spec") account for 80 percent of the defects, as indicated by the point on the line above "Out of Spec" on the right-hand axis.

Further analysis can decompose the data and plot additional Pareto charts. For example, missing parts may come from several suppliers, as seen in the following table and used for the drill-down Pareto chart in Figure 4.6:

Missing parts by the supplier	Count
Ace Co.	9
Long Supply Co.	6
Foreign Source	3
New Alternate Co.	2
Best Supply Co.	1

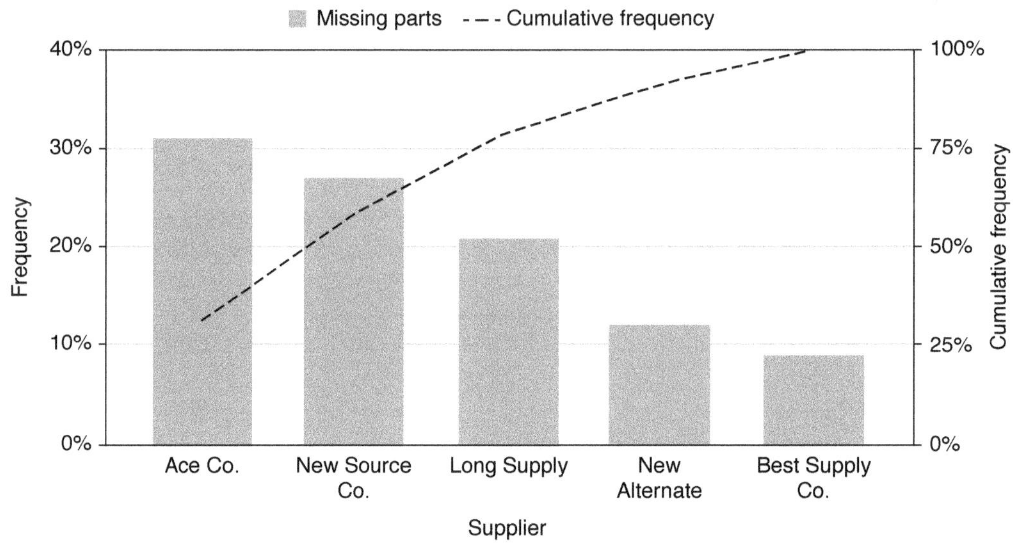

Figure 4.6 Drill-down Pareto chart.

Cause-and-Effect Analysis: Understanding Why the Costs Exist

Brainstorming is a good approach for collecting potential causes of problems when the team is experienced. Sources of team focus are established earlier with the cause-and-effect chart, also known as an *Ishikawa diagram* or, more commonly, a *fishbone chart*. Popularized by Kaoru Ishikawa in the 1960s, this brainstorming framework (Figure 4.7) helps a team identify potential causes of a problem.

1. It starts with describing the problem so that everyone understands the scope. Write this in a box in front of the group at the right side and draw a box around it. Leave most of the space blank.
2. Draw an arrow left to right ending at the problem box.

3. Choose broad categories of causes that make sense to the group. There are several standard sets of categories:
   - Eight *m*'s: manpower, machine, material, method, measurement, Mother Nature (environment), management, maintenance
   - Five *s*'s for service: surroundings, suppliers, systems, skill, safety
   - Four *p*'s: product, place, people, performance
4. Draw and label boxes for the categories above and below the main arrow and place diagonal arrows pointing to the main arrow.
5. The group brainstorms one category at a time, leaving the "people" or "manpower" one for last, to avoid blaming.

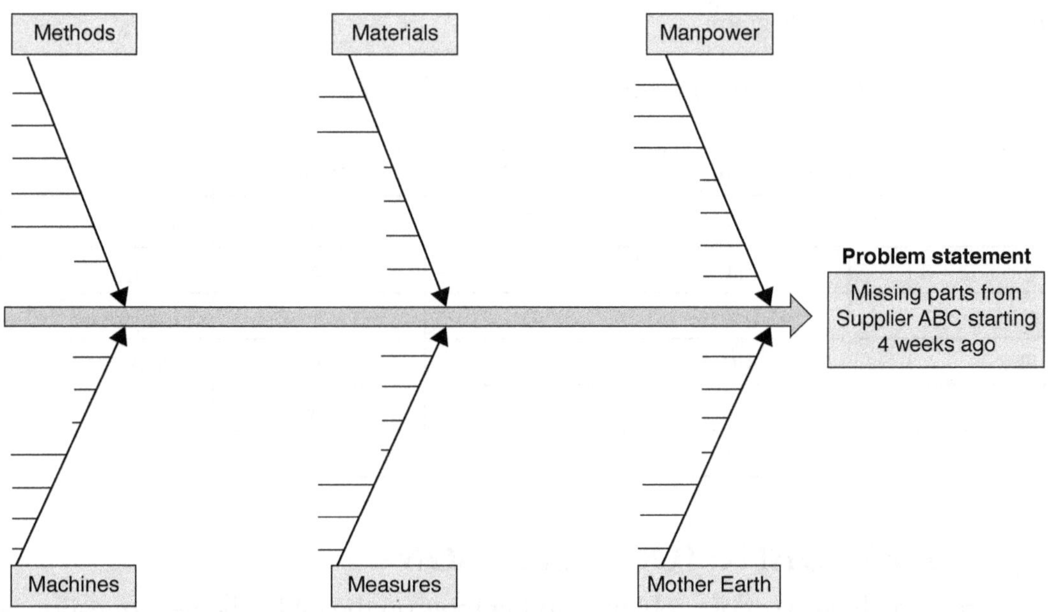

Figure 4.7 Cause-and-effect chart.

Relations Diagram: Seeing How Causes Interact

Once causes are identified, they may need to be sorted. One helpful way to sort causes is to determine which ones depend on others. The relations diagram shown in Figure 4.8 is an example (also called the *interrelationship digraph*).

1. Causes are arranged in a circle, leaving space in the middle.
2. Considering each cause, draw arrows from that cause to any and all others that are driven by it.
3. Count arrows going into each cause and arrows going out from each cause.

4. Create a proportion of ins and outs.
5. Sort causes in order of their proportions. No arrows going in: It is a final cause. All arrows going out: It is a trigger cause.

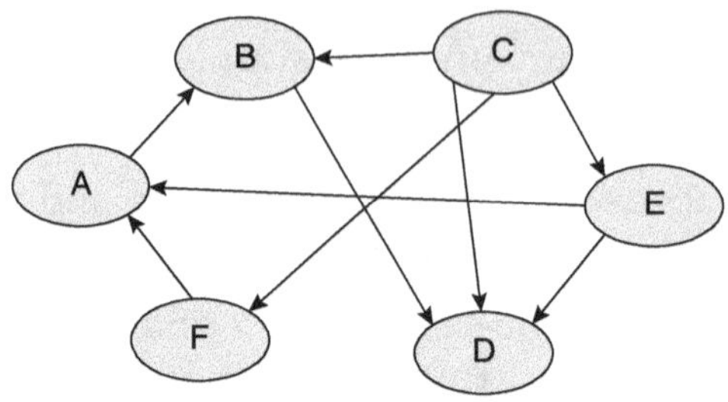

Figure 4.8 Relations diagram.

In this diagram, the in-out proportions are as follows:

	In:Out	Result	Order	
A	2:1	2	4	Middle
B	2:1	2	4	Middle
C	0:4	0	1	First driver
D	3:0	3	5	Driven
E	1:2	0.5	2	Middle
F	1:1	1	3	Middle

WHAT TOOL WHEN: TOOL SELECTION

Once you have a set of problem-solving tools on the table, the next question is obvious: Do we use all of them, or just one? In control work, the answer is almost always only what is needed. The tools we just covered are not the only options, but they deliver a substantial impact for the effort. They are meant to help you move from a drift signal to a decision without turning every issue into a full-scale project.

The flow begins with a drift trigger and moves through three phases. Phase 1 is a quick reality check and category trending, just enough information to separate real change from accounting noise, timing effects, or mix shifts and to see how the COQ categories are actually behaving. Phase 2 is where you choose a tool. You name the pattern (global shift, category trade-off, or localized spike) and then pick the lightest tool that fits. Phase 3 closes the loop. You test what you think is driving the change, verify stability on the trends with lag in mind, and lock the gain into standard work and routine management. You may use a different system in your organization, and that's fine. The utility is not the exact sequence; it is ensuring that drift turns into action and that action turns into sustained control.

Trigger: Detect the drift

- The drift flag originates from a COQ trend move *or* a leading quality signal (with COQ confirmation to follow).

Phase 1: Initial analysis (quick + trend)

- *Quick reality check*
    - Any changes in definitions / coding (bucket rules)?
    - Any changes in rates / costing rules (burden / standards / annual recosting)?
    - Any timing / lag effects (late postings, accrual timing, month-end effects)?
    - Any mix changes (product / customer / supplier mix, volume swings)?
    - Any one-time events (shutdowns, turnover / retirements, major rework campaigns)?
- *Trend the categories (not just total)*
    - Plot *investment, monitoring, waste, and waste + downstream consequences* over time.
    - Mark intervention dates and note expected lag.

Phase 2: Classify the pattern (name the problem)

- *Global shift (most categories move together)*
    - Meaning: system-wide change or measurement / recordkeeping change.
    - Ask: Are there cultural shifts or changes to policy to consider?
    - Name tool: Fishbone (unclear) / relations diagram (cross-functional + tangled).
- *Category trade-off (up while another is down)*
    - Meaning: costs shifted upstream / downstream or moved between categories.
    - Ask: "Planned?" and "Favorable end-to-end (waste + downstream consequences down)?"
    - Name tool: Relations diagram (interdependencies) / timeline analysis (sequence of events diagram).
- Localized spike *(one bucket/area jumps; others stable)*
    - Meaning: concentrated driver in a product / process / supplier / customer segment.
    - First move: stratify to find where it lives.
    - Name tool: Pareto → then 5 whys (apparent direct cause) or Ishikawa / fishbone (unclear / unknown).

Phase 3: Close the loop (test + verify + lock-in)

- *Test the likely drivers*
    - Confirm with targeted checks (data pull, audit, short trial) before broad corrective action.
- *Verify on trends (expect lag)*
    - Look for return to baseline band and sustained stability, not a one-period dip.
- *Lock in the gain*
    - Update standard work, training, checks/audits, ownership, and review cadence, so control is routine.

RISK-BASED CONTROL: PREVENTING THE NEXT LOSS

There are two ways to consider risk: The first one is *enterprise risk*. This is generally handled using ISO 31000, which considers all broad-based systematic risks an organization may face. The other way to assess risk is to use what ISO 9001:2015 calls risk-based thinking. These risks are more day-to-day and decision-by-decision considerations.

In a quality context, the primary tool for assessing risk is failure modes and effects analysis (FMEA). This tool comes in several varieties and can be applied to product design, process design and operation, broad-based systems, and other areas. In using process-FMEA, you first have to identify the steps in your process. Then a team brainstorms all the ways each step might fail, along with the severity of each failure, its frequency, and the likelihood of detecting it before its effects are observed. (The following describes a general approach to FMEA more similar to the 2008 automotive approach, as opposed to the newer 2019 Automotive Industry Action Group (AIAG) & Verband der Automobilindustrie (VDA) approach. The older approach is sufficient and similar, so we use it here. The new variety may be used if your organization has the architecture and process expertise to implement it. These three scores (severity, occurrence, detection) are measured on a team consensus score from 1 to 10. The three scores for each failure mode at each step are multiplied together, producing a 1-to-1,000-point scale of risk for that failure mode.

Sorting by the 1,000-point scale surfaces the highest-risk items. Action plans are developed for all major failure modes. COQ may be added to an FMEA by using cost estimates for the riskiest areas based on their severity, then multiplying them by their occurrence frequency. This yields an expected cost for each failure mode, and the corrective actions identified in the FMEA are then estimated accordingly. Ongoing monitoring costs may be added as another step in this analysis.

The return on investment for each corrective action can be compared with the expected value of failure (the cost of failure multiplied by its probability), enabling a financial appraisal of the corrective actions. It is important to remember that a corrective action may affect more than one failure mode; in this case, the investment in the corrective action should be compared with the combined expected values of the linked failure modes.

In the example shown in Figure 4.9a, an assembly step has identified a failure mode: parts installed upside down. Looking at the severity, occurrence, and detection for this example, the overall risk priority number is 288. Compared to other risks in the same analysis, this was identified as a high-priority area for solution.

Process Step	Function	Failure Mode(s)	Effects	Sev	Cause of Failure	Prevention Controls	Occ	Detection Controls	Det	RPN= SxOxD
Assembly	Complete sub-assembly attach-ments	Part installed upside down	Total functional failure	9	missing instructions and poor training	None	4	Infrequent visual inspection by the operator	8	288

Figure 4.9a FMEA example.

The solutions team (see Figure 4.9b) identified two actions: The first was to include a qualification test for operators before assigning them to assembly. This reduced the occurrence value from 4 to 2. The second action added a brightly colored mark at the top of the part, which could be installed upside down, so the operator would much more easily detect an upside-down installation. This reduced the detection score from 8 to 6. The risk priority number declined from 288 to 108. This was deemed satisfactory as a risk-reduction activity, but three additional columns were added to the FMEA spreadsheet to assess the economic benefits of these changes. The investment in a training program, plus the partial modification to add the color indicator, was estimated at $2,500. Using quality cost records, the company determined that, over the course of a year, this problem cost $14,000. Dividing 14,000 by 12 yields the monthly waste cost, and dividing the investment by this monthly waste cost yields a payback of 2.1 months. This kind of information indicates the financial return from the FMEA analysis.

Possible Occurrence Action(S)	Possible Detection Action(S)	New OCC	New Det	Revised RPN	Investment	Annual Waste Reduction	Payback (Months)
Operator qualification test before assignment	Color indicator on the part top	2	6	108	$2,500	$14,000	2.1

Figure 4.9b FMEA example.

CONCLUSION

Improvement gets the applause. Control keeps the trophy on the shelf. A COQ system gives you the numbers to spot drift early, justify corrective action with real economics, and keep teams aligned on what matters after the "project" ends. When you pair day-to-day quality measures with trend-based COQ reporting, you get both speed and stability: current signals to act now and cost confirmation to prove the gains are sticking. The organizations that sustain results are not the ones that fix the most problems. They are the ones who build control into the way work runs.

TOOL: COQ-BASED CONTROL CHECKLIST

1. *Is the control target clear?*
   What outcome must not drift, and where in the value stream?
2. *Do we have a stable baseline?*
   What does "normal" look like for total COQ and for each COQ category?
3. *Are we sure the signal is real?*
   Any changes in coding rules, costing rates, ledger timing, product mix, or recosting that could explain it?
4. *Do we understand the time lag?*
   Which leading operational measures should move before COQ moves, and are they changing too?
5. *Is drift global, a category shift, or localized?*
   Total up COQ everywhere versus trade-off between categories versus one bucket/area spiking.
   Have we trended each category separately over time?
   Not just totals—each category's trend line, with intervention dates marked.
6. *Have we stratified to find where it lives?*
   By product family, line/shift, plant, supplier, customer segment, region—whatever reveals "where."
7. *Do we have trigger rules and owners?*
   What threshold counts as drift, who is notified, and who must respond?
8. *Is there a defined reaction plan?*
   Containment → cause analysis → corrective action → verification → lock-in.
9. *Is the economics test explicit?*
   Cost of action versus cost of inaction/payback, using COQ.
10. *Did we verify stability after the fix?*
    Evidence of "back in baseline band" for a sustained period, considering lag.
11. *Did we lock it in?*
    Standard work, training, checks/audits, and clear accountability are updated to ensure the gain is sustained.

KEY TAKEAWAYS

- *Control is the missing link.* Without it, improvements regress, and costs quietly creep back in.
- *COQ makes drift visible.* Baselines, thresholds, and routine trend reviews turn surprises into early warnings.
- *Corrective action should pay back.* Use COQ to compare the one-time cost of a fix to the ongoing cost of doing nothing.
- *Ownership must be cross-functional.* Many quality costs originate upstream, so control must span engineering, operations, documentation, finance, and HR.
- *Trend charts tell the truth over time.* Use long-range trends for strategy and short-range trends for local control and accountability.
- *Expect time lags.* Quality metrics are current; cost data arrives later. Use both and annotate charts with intervention dates.
- *Use the right tools to turn cost into action.* Pareto identifies the vital few; fishbone and relationship diagrams organize causes; and FMEA adds risk through a financial lens.
- *COQ strengthens suppliers and small businesses.* Control reduces repeat errors when resources are tight, and supplier drift can become expensive quickly.

Chapter 5

Quality Improvement and Reducing Costs

Continuous improvement is a core principle in quality and just as important in managing the COQ. This chapter shifts the focus from implementation and stabilization to improvement, specifically how to apply COQ to reduce costs and enhance performance.

While improvement is often presented as a clean step-by-step process, real-world applications rarely work that way. Teams move through cycles, test ideas, revisit assumptions, and adapt as they go. The structure of this chapter reflects that reality. It follows the plan-do-check-act (PDCA) cycle, not as a rigid sequence but as a practical framework for navigating change.

If you're the kind of reader who skips ahead to find the "good stuff," you're in the right place. This chapter offers hands-on guidance, practical methods for applying COQ analysis, and tools to support structured problem-solving and cost-reduction efforts. Nevertheless, if you haven't reviewed Chapters 3 and 4 on planning and control, consider circling back. Strong improvement efforts are grounded in solid planning and effective control. Just as with the Juran trilogy, the three work best as a system.

You'll find techniques here to help reduce total quality costs and improve profitability. We'll also explore how COQ aligns with other improvement philosophies, such as lean and Six Sigma, and how it connects long-term strategy with daily operations.

Although many of the examples come from manufacturing, the principles apply equally well to service industries, such as banking, healthcare, insurance, and retail. Later chapters address those contexts more directly, but the fundamentals remain the same: Improving quality reduces total cost.

It's also worth remembering that quality costs aren't limited to operations. They span the entire life cycle—marketing, design, procurement, production, delivery, and service. As a result, any serious COQ initiative must be cross-functional. And while a dollar saved is often the visible goal, the customer is the real goal.

Why Improvement Drives Profit

It is a fact, though too often overlooked, that every dollar saved in the total COQ translates directly into a dollar of pretax earnings. The implication is that a dollar-for-dollar cost reduction has a greater impact on the bottom line than an increase in sales. It is also a fact that quality improvements and cost reductions cannot be achieved by declaration alone. They must be earned through structured, disciplined problem-solving.

The first step in this process is identifying problems, which, in the context of COQ, we define as areas of high quality cost. Every such problem represents an opportunity to improve profitability.

It's widely understood that many functions influence quality, including marketing, design, purchasing, and service. Still, too often, quality initiatives remain narrowly focused on production alone. In other unfortunate cases, quality efforts have been limited to maintaining the status quo (control) rather than pursuing continuous improvement. As a result, performance has declined in many areas because control alone rarely proves 100 percent effective.

Improving quality, much like managing costs, must be a company-wide effort. Everyone supports the idea in principle, but visible management commitment and a formal program to drive improvement are key to success.

Each function—marketing, design, purchasing, accounting, manufacturing, and quality assurance—must help prevent defects and reduce waste. Tools are necessary to help identify quality-related problems, uncover their root causes, and eliminate them.

The message is clear: Quality improvement leads directly to cost improvement. Designing and building a product correctly the first time always costs less. Solving problems at the source results in measurable, sustained savings. To realize these savings, quality performance must be elevated and maintained.

Figure 5.1 illustrates how quality cost analysis bridges the gap between the tactical efforts of a prevention-oriented quality program and the financial metrics used by leadership to measure performance—specifically, the profit and loss statement. The figure shows the flow of COQ data from operational quality assurance activities to the total COQ and ultimately to financial performance.

Every dollar saved through improved quality has a direct and traceable impact on profitability.

Quality Costs and the Profit Center

Quality costs for a profit center are generated across a range of organizational activities. Figure 5.2 illustrates how costs from all functional departments are collected and compiled into a comprehensive quality cost analysis for the entire profit center.

It's important to note that many organizations are structured as cost centers, particularly when production and sales activities are functionally separated. This doesn't imply that only manufacturing or service operations are viewed as costs; rather, it reflects how responsibilities are divided for financial tracking and accountability.

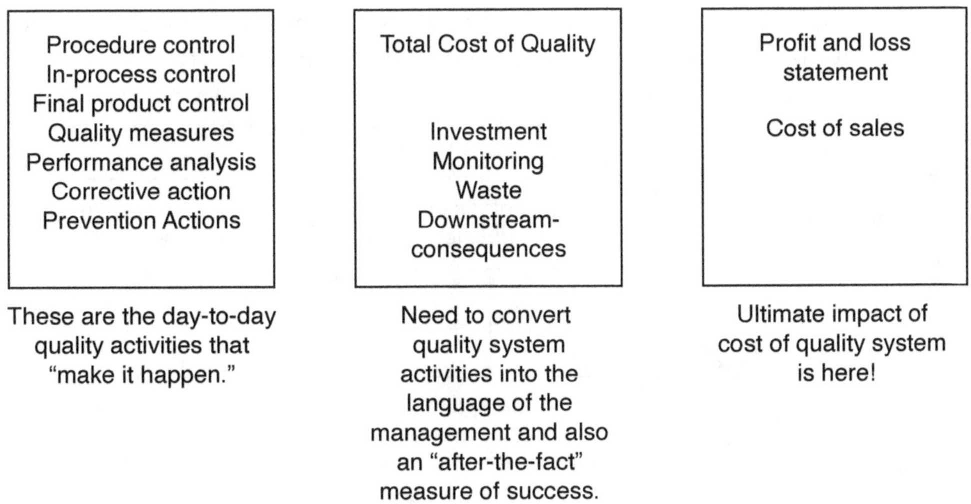

Figure 5.1 Quality cost system.

In such structures, practitioners may have less influence or visibility into such functions as marketing when applying COQ principles. Nonetheless, when necessary, organizational silos must be broken down. A clearly defined charter or shared understanding of roles, responsibilities, and authorities can go a long way in enabling managers to focus their efforts where they're most impactful for the organization's specific structure and goals.

Looking at the example costs in Figure 5.2, it becomes clear that quality costs are incurred across all primary functions—not just production. Problems can originate anywhere. To identify and prioritize areas for improvement, careful cross-functional analysis is required.

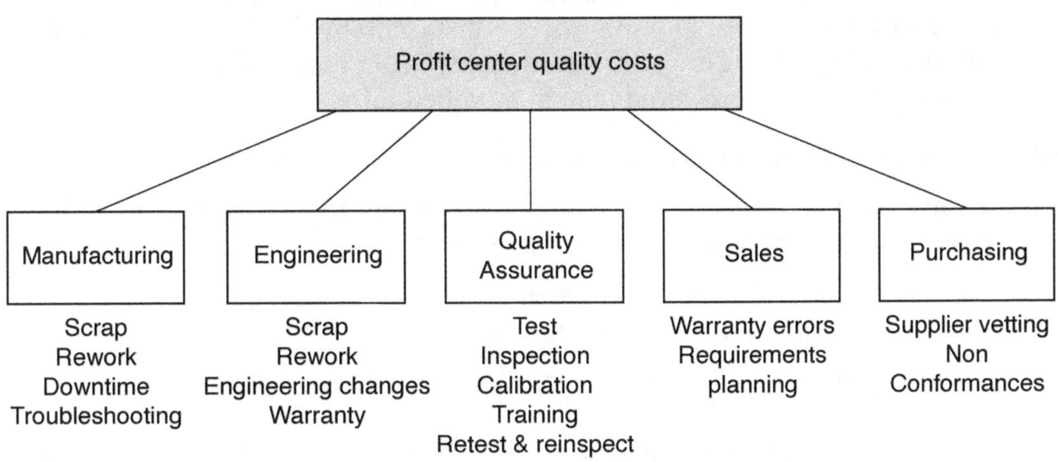

Figure 5.2 Profit center quality costs.

Once these critical areas are identified, programs should be developed in collaboration with all relevant departments to ensure a practical, collaborative approach to reducing aggregate quality costs. Figure 5.3 shows the relationship between the overall strategic program and the quality program.

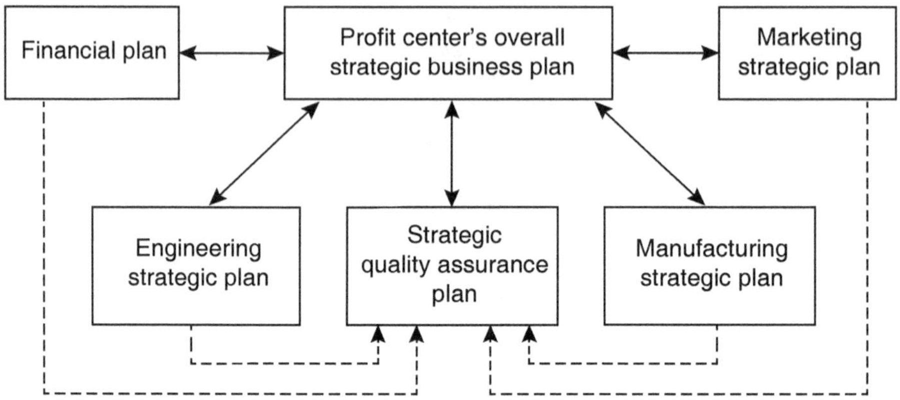

Figure 5.3 Profit center's overall strategic business plan.

PLAN: START WITH THE CUSTOMER, NOT THE METRIC

Customer Satisfaction Makes the Case

Quality cost techniques can be used to justify efforts to improve customer satisfaction. The monetary gain a company receives from satisfied customers is quantifiable. A satisfied customer is more willing to repurchase from a company. The payoff is an increase in repeat sales and, eventually, an accompanying increase in cash flow. There is no better illustration of this principle than William Winchell's 1996 article "Driving Buyer Satisfaction by Quality Cost," from the 50th Annual Quality Congress Transactions. Being indescribable, it is, as with previous editions, reproduced here in modified form.

Why Customer Satisfaction Matters to COQ

To have an effective continuous improvement effort, a company must have specific values and beliefs, including the following:

1. A focus on buyers and their needs
2. A recognition that the buyer determines quality
3. An awareness that problem prevention is of primary importance

In a continuous improvement environment, quality cost techniques must focus on the buyer. The approach described here supports constant improvement efforts. It recognizes that buyer perceptions form the benchmark of quality. It also justifies changes that can prevent problems that affect the buyer.

What Buyer Satisfaction Really Tells You

Like beauty, quality is in the eye of the beholder. In this case, the beholder is the buyer. This perception of a buyer is of great importance to a company. A satisfied customer is more likely to repurchase from a company in the future. This relationship is shown in Figure 5.4.

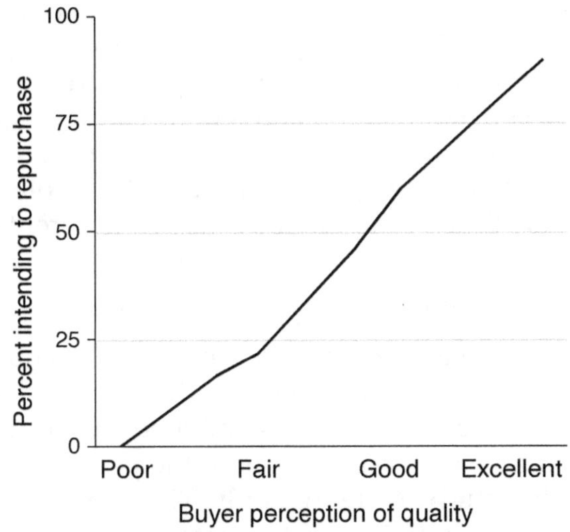

Figure 5.4 Repurchase intentions based on perception of quality.

Source: Based on data from marketing study. J. M. Juran and F. M. Gryna, *Quality Planning and Analysis*, 3rd ed. (McGraw-Hill, 1993), 87.

What the Organization Gains

The potential monetary gain, or payoff, a company experiences from satisfied customers is often significant. A satisfied customer is much more likely to repurchase. Because this payoff may be in the future rather than immediate, the company must have a long-term outlook.

In contrast, the penalties for buyer dissatisfaction are often immediate and severe. An unhappy buyer could tell as many as thirty-five people about the problem, and this secondary effect may have a severe impact on future sales.

A gain in contribution margin is the payoff a company experiences by improving customer satisfaction. Gain in contribution margin is defined as cash flow from the incremental gain in revenue minus the incremental increase in cost due to additional repeat sales from more satisfied customers.

The following steps are suggested to calculate the gain in contribution margin resulting from a projected increase in buyer satisfaction:

1. *Find the relationship between buyers' perception of quality and repurchase intentions.*

Determining the relationship between buyers' perceptions of quality and repurchase intentions requires a marketing study of the current buyer base.

The buyers chosen for the survey should be those likely to repurchase soon. Ask buyers to rate the product's quality, then whether they would repurchase a similar product. Figure 5.5 shows the results of such a study.

2. Determine when repurchases are likely to be made.

For most companies, repurchases occur near the end of a product's life cycle, typically well beyond the warranty period. For some companies, sales records may contain this information. For others, a marketing study of the current customer base may be required. Ask buyers when they expect to repurchase a similar product. Figure 5.5 is a profile of buyers' repurchase intentions. All buyers initially purchased the product at time "0." Note that each company and each product within a company may have a different profile.

3. Assess buyers' current perceptions of product quality.

Assessing buyers' current perceptions of the product's quality also requires a marketing study of the customer base. To reduce effort, the marketing studies suggested in these three steps may be combined into a single survey. For this step, ask buyers to rate their feelings about the product. Figure 5.6 illustrates the results of such a study. Note that only 30 percent of customers rated the product good or better. That 30 percent leaves room for significant improvement in customer satisfaction.

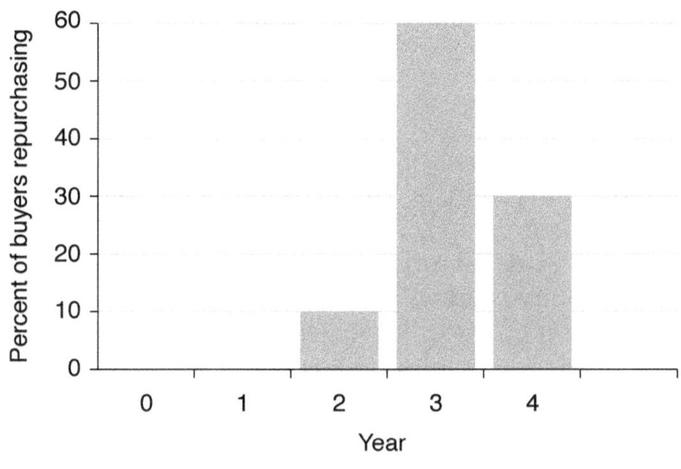

Figure 5.5 Profile of when buyers in year 0 will repurchase.

4. Project the desired change in customer perceptions of quality before making improvements.

Much judgment is required as to what level of satisfaction should be sought. The payoff for a company depends on the size of the projected increase. For example, a larger increase will yield greater monetary gain from repeat sales. On the other hand, a larger increase will be more challenging to obtain. Aiding in this judgment is knowledge of the current level of customer satisfaction (as found in Step 3). Comparing customer satisfaction with that of owners of competitive products may also be helpful, and a marketing study is necessary. Figure 5.6 also shows projected perceptions of quality following a proposed increase in customer satisfaction. Note that customer perceptions are projected to increase by one notch following the improvements.

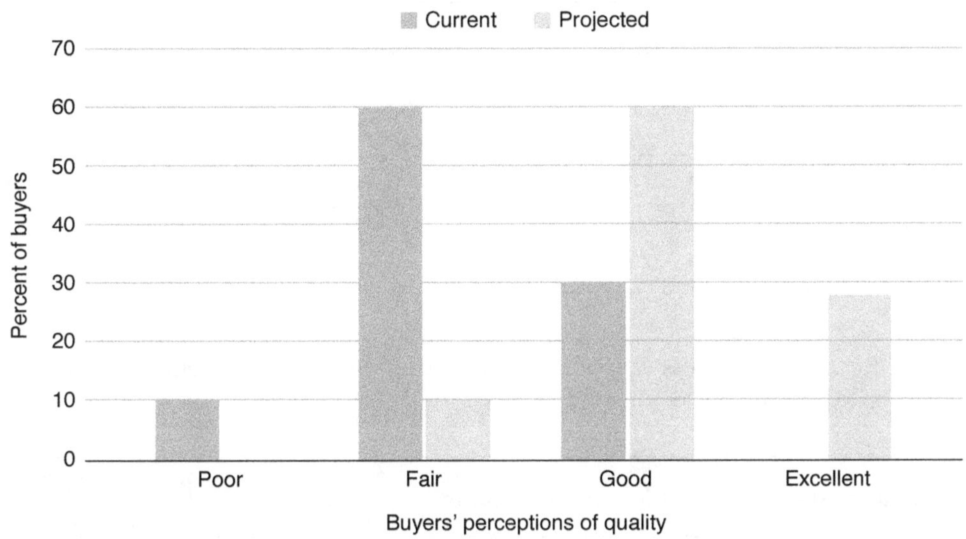

Figure 5.6 Buyers' current perceptions of quality and projected perceptions after the proposed increase in customer satisfaction.

5. *Calculate the percentage of buyers willing to repurchase under current levels of customer satisfaction.*

The calculation of current customer satisfaction levels, as measured by the "composite % willing to repurchase," is shown in Table 5.1.

The product of columns 2 and 3 is entered in the fourth column ("Adjusted % willing to repurchase") after converting from necessary percentages to decimals. The total of the entries in the fourth column comprises the "Composite % willing to repurchase."

This measure of the current customer satisfaction level is a weighted average of the perception data. Table 5.1 shows that only 29.4 percent of buyers are currently willing to repurchase.

6. *Calculate the percent of buyers willing to repurchase under the projected levels of customer satisfaction.*

A similar calculation can be made for projected customer satisfaction levels (Table 5.2). The data sources and calculations are identical to those described in step 5. The projected customer satisfaction measure ("Composite % willing to repurchase") shows that 66.6 percent are now likely to repurchase.

Table 5.1 Buyer's adjusted willingness to repurchase with current levels of customer satisfaction.

Buyers' perception	% willing to repurchase	% of buyers	Adjusted % willing to repurchase
Poor	0	10	0
Fair	18	60	10.8
Good	62	30	18.6
Excellent	92	0	0
Composite % willing to repurchase			29.4

Table 5.2. Buyers' willingness to repurchase with proposed improvement in customer satisfaction.

Buyers' perception	% willing to repurchase	% of buyers	Adjusted % willing to repurchase
Poor	0	0	0
Fair	18	10	1.8
Good	62	60	37.2
Excellent	92	30	27.6
Composite % Willing to Repurchase		100	66.6

7. *Calculate the gain in percent of buyers willing to repurchase if projected levels of customer satisfaction are obtained.*

The increase in buyers willing to repurchase is 37.2 percent, the difference between "Composite % willing to repurchase" in step 6 (66.6 percent) and in step 5 (29.4 percent). In other words, repeat sales from the projected increase in buyer satisfaction more than doubled.

8. *Calculate the gain in contribution margin for each year repurchases are made if projected levels of customer satisfaction are obtained.*

Table 5.3 shows the increase in contribution margin if projected customer satisfaction levels are achieved.

Table 5.3 Calculated gain in contribution margin due to gain in repeat sales through proposed improvement in customer satisfaction.

Year	% buyers repurchasing	Gain in % willing to repurchase	Volume bought year 0	Contr. Margin $/unit	Gain total contr. margin
0	0				
1	0				
2	10	37.2	200,000	4.75	35,340
3	60	37.2	200,000	4.75	212,040
4	30	37.2	200,000	4.75	106,020

The product of columns 2 through 5 for each row is entered in column 6 ("Gain total contr. margin"). Note that this calculation requires converting percentages to decimals. The payoff to the company, or the gain in total contribution margin, is shown for years 2 through 4, representing when buyers will make repurchases. However, investment that improves buyer satisfaction would mostly be made in year 0—the year the product is built and sold. For this example, the payoff from repeat sales and additional cash flow will be realized over two to four years after the product is built and sold. Customer satisfaction often requires a long-term outlook to justify investments.

9. *Find the present worth in year 0 of the gain in contribution margin for future years.*

Investments to improve customer satisfaction must be made primarily in year 0 of the product's life, before shipping. Therefore, the present worth (year 0) of the benefits listed in Table 5.3 for years 2 through 4 must be calculated. Time value of money (TVM) conversions will allow an "apples-to-apples" comparison of the benefits to the cost of making the necessary changes required in year 0. Table 5.4 shows this relationship. Economic principles of net present value (NPV) can be used to make this calculation. For example, assuming a 10 percent compound interest factor, the calculation is as shown in Table 5.4.

Table 5.4 Calculate present value.

Present worth = sum (future amount × compound amount factor)						
(Year 2)	=	$35,340	×	0.8264	=	$29,205
(Year 3)	=	$212,040	×	0.7513	=	$159,306
(Year 4)	=	$106,020	×	0.6830	=	$72,412
(Year 0)	=					$260,923

More information on NPV and other TVM concepts is found in Appendix A.

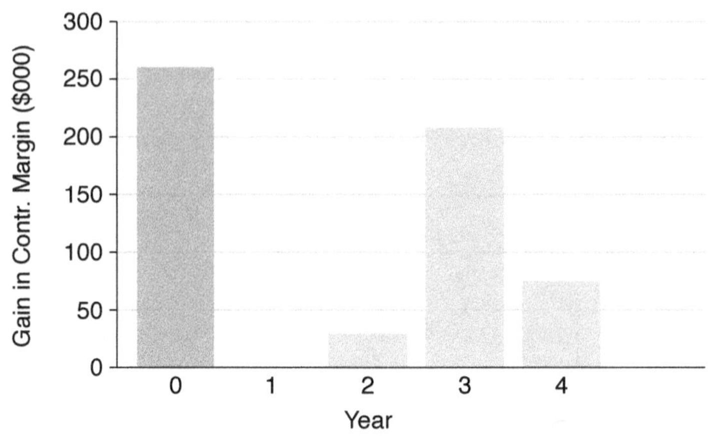

Figure 5.7 Gain in contribution margin. Present worth in year 0 of gain in contribution margin through increased repeat sales in years 2–4. The proposed rise in customer satisfaction is expected to drive additional repeat sales. A compound interest factor of 10 percent was assumed.

Justification: Making the Case in Dollars

More than $260,000 has been identified by the new approach for justifying changes that improve customer satisfaction. This estimate is in addition to funds that traditional quality cost techniques may locate. Traditional quality cost techniques may identify other funds supporting changes leading to improved customer satisfaction through reductions in several areas:

- *Warranty claims.* The total cost of claims paid to the buyer or user, after acceptance, to cover expenses, including such repair costs as removing defective hardware from a system or cleaning costs due to a food or chemical service accident. When a price reduction is negotiated in place of a warranty, the value of the reduction should be recorded.
- *Liability costs.* Company-paid costs due to liability claims, including the cost of product or service liability insurance.
- *Penalties.* The cost of any penalties incurred because of less than full product or service performance (as required by contracts with buyers or by government rules and regulations).
- *Buyer/user goodwill.* Costs incurred above normal selling costs for buyers or users who are not fully satisfied with the quality of the delivered product or service.

Investment: Turning Insight into Targeted Investment

Over time, the definitions of quality costs were adapted to apply to work beyond manufacturing. The cost definitions now correspond to activities in the quality system for producing a product or service. The definitions begin in marketing and extend to other functions related to product development and production. In view of this, quality cost definitions are used to show where possible investments, before or after product realization, can be made to upgrade customer satisfaction.

About 80 percent of buyer satisfaction depends on whether a product meets buyers' actual needs. Improvements in assessing buyer needs and validating that they are met leads to greater buyer satisfaction. Improvements may be in the following quality activities:

- *Marketing/buyer/user.* Costs incurred in the collection and ongoing evaluation of buyer and user quality needs and perceptions (including feedback on reliability and performance) that affect their satisfaction with the company's products or services. Activities may include
  - marketing research,
  - buyer / user perception surveys / clinics, and
  - contract / document review.
- *Product/service design development.* Costs incurred to translate buyer and user needs into reliable quality standards and requirements and manage the quality of new product or service developments before the release of authorized documentation for initial production. This includes
  - design quality progress reviews,
  - design support activities,
  - product design-qualification test, and
  - service design-qualification field trials.
- *Special product evaluations.* Includes life testing and environmental and reliability tests performed on production units.

The remaining 20 percent of buyer satisfaction perception is influenced by how a company handles complaints and resolves problems. The identified funds could also support desirable changes for these activities. Activities may include the following:

- *Evaluation of field stock and spare parts.* Includes cost of evaluation testing or inspection of field stock resulting from engineering changes, storage time (excessive shelf life), or other suspected problems.
- *Complaint investigations of buyer or user service.* The total cost of investigating, resolving, and responding to individual buyer or user complaints or inquiries, including necessary field service.
- *Retrofit costs.* Costs to modify or update products or field service facilities to a new design change level due to major redesigns resulting from design deficiencies.

Although not a quality cost, improving customer satisfaction may also require investing in product upgrades or process improvements.

Key Takeaways from Customer-Driven COQ

Looking at how customer satisfaction links to quality cost measurement, Winchell (1996) concludes the following:

1. Quantifying the monetary gain a company derives from customer satisfaction provides strong justification for improving factors that affect it.
2. Companies using this approach must have a long-range outlook. The monetary gain occurs in the future through repurchases.
3. Future monetary gains must be discounted to present value to determine when to invest in improvements. Calculations can be done using engineering economics principles, creating an "apples-to-apples" comparison between investment costs and expected benefits or monetary gain.
4. The suggested approach expands on costing techniques by providing a customer-focused analysis. Customer-focused analysis is more aligned with continuous improvement, lean, and Six Sigma approaches.

From Cost Visibility to Targeted Improvement

With customer focus well established, organizations can shift to an internal focus by creating a strategic quality plan that outlines management's commitment to quality and to reducing quality costs. COQ plays a role here again by helping identify key areas that are strong candidates for improvement. When the highest-cost regions are analyzed in more detail, numerous improvement opportunities typically emerge.

For example, high downstream costs, such as warranty claims, can trigger prioritization of customer failure issues for investigation. Analysis may reveal opportunities for targeted actions in product design, process control, or inspection planning, ultimately aimed at eliminating the highest-cost problems. Regardless of which COQ category is responsible for the high cost, simply identifying it should help focus attention and drive action to reduce it.

Today, organizations are often highly specialized, and most leaders understand that there are no generic solutions to quality issues. High-cost areas are not resolved through accreditation, certifications, organizational restructuring, management fads, or even COQ analysis alone. Tools and tactical problem-solving drive success. The complementary nature of COQ to tactical, practical problem-solving is why COQ is a perennial tool for linking strategic management from the boardroom to the shop floor. COQ data serves as a spotlight, identifying areas of concern but not resolving them. Once a high-cost problem is identified, it must be examined in detail, and appropriate, often customized, corrective actions must be taken. Identification and action can, of course, be united in a single cohesive program, but the fact remains that cost reduction is achieved one problem at a time.

Because of the downstream consequences, it can be tempting to increase monitoring to catch problems earlier. However, this may only shift costs from external failures to waste, such as rework or scrap, and increase inspection workload. While such a shift may be acceptable as a short-term or stopgap strategy, it doesn't address the root cause.

Similarly, some may strengthen product engineering efforts across the board to prevent defects. But generalized investment activities may not be efficient. The most effective investment actions target specific, identified issues highlighted through COQ analysis.

Put simply, successful quality improvement and COQ reduction efforts depend on understanding cause-and-effect relationships. In this regard, the study of quality costs remains one of the most effective tools available to management for visualizing and acting on those relationships.

Identifying High-Impact Cost Drivers

When quality cost data is presented to managers for the first time, the most common questions are "How much should these costs be?" or "How do we compare to other organizations or products?"

Unfortunately, it is rarely practical or even meaningful to establish absolute standards for quality cost comparisons. A quality cost system must be tailored to the specific needs, context, and strategy of the individual organization. While industry benchmarking can offer valuable insights, it should be treated as reference data, not as a prescriptive target. Even within the same industry, companies differ in their structures, processes, goals, and market approaches. If every competitor were pursuing the market in the same way, it would be an unproductive race to the lowest price and lowest margin.

Instead, organizations must rely on experienced judgment, ideally collective and informed by a deep or at least evolving understanding of the business and its operations. This perspective is essential for identifying meaningful trends and providing objective evidence to inform decisions on where to focus quality assurance efforts to maximize return on investment.

While the desire to seek universal guidelines or external benchmarks is understandable and even encouraged by such practices as benchmarking, it can be risky. Overreliance on external comparisons can lead to a scorecard mentality, where COQ becomes just another metric to track rather than a management tool for driving meaningful improvement. (Note: This type of "scorecard" should not be confused with the balanced scorecard methodology discussed in Chapter 10.)

The limitations of absolute quality cost comparisons become even clearer when you consider the following:

- significant variation in how companies define and capture quality cost data;
- differences in product complexity, life cycle, and process stability;
- varied production volumes, market dynamics, customer expectations, and organizational maturity; and
- the mismatch between the timing of revenue recognition (e.g., net sales billed) and when quality costs are actually incurred.

This last point is crucial in industries with volatile sales volumes, shifting product mixes, or changing business models (e.g., service vs. leasing). In such cases, the commonly used metric, COQ as a percentage of net sales, may not offer an accurate or fair picture.

Given these realities, organizations are far better served by focusing on internal analysis. Instead of comparing numbers to external entities, analyze which areas contribute most to your own quality costs, and pursue targeted corrective actions based on those insights.

Just as a doctor wouldn't prescribe treatment based solely on another patient's experience, your approach to quality cost management must be tailored to the unique needs of your organization. The tools and techniques for analyzing quality costs are as diverse as those used to solve other business problems. They range from simple charts and visualizations to more complex mathematical models. If your organization reaches a point where advanced tools are necessary, the capacity to use them can be developed over time. However, in most cases, a thoughtful, committed effort with simple tools will be far more effective than a rushed, superficial attempt at something overly complex.

Planning Summary: Key Steps for Effective Quality and Cost Improvement

- Recognize and organize quality-related costs to gain insight into their magnitude, contributing elements, and trends.
- Analyze quality performance to identify major problem areas and assess product line or departmental performance.
- Implement practical corrective actions and targeted cost-reduction programs.
- Evaluate the impact of those actions to ensure they achieve the desired results.
- Prioritize efforts for maximum financial return and optimal use of personnel.
- Budget quality activities in alignment with defined performance objectives.
- Coordinate actions across investment, monitoring, waste, and downstream consequence categories to preserve visibility into cause-and-effect dynamics.

DO: A PRACTICAL COQ PROBLEM-SOLVING FRAMEWORK

Planning is an indispensable catalyst for continuous improvement, but planning without action is useless. COQ techniques support a bias for action and are especially valuable for implementing quality improvement through teams, helping them quantify the financial impact of their projects, both for internal motivation and to communicate value to management.

While the specific problem-solving framework used—whether DMAIC, The 8 Dimensions (8D), Plan-Do-Check-Act (PDCA), or a seven-step approach—is less important, what matters is that teams follow a structured and disciplined process that addresses all critical improvement stages.

The following example outlines a ten-step format that integrates quality cost principles into the problem-solving process. It originates from a service organization, but its structure is equally applicable to manufacturing environments. The key takeaway is that linking financial insights with root cause analysis and corrective action planning gives teams a more powerful way to demonstrate and deliver results.

Step 1. Form or Select Cross-Functional Teams

The most effective way to secure management support for quality improvement efforts is to demonstrate how team-based COQ techniques can positively impact the organization's bottom line.

By focusing on key business processes, such as order administration, new product development, sales, service, or manufacturing, teams can use quality cost measurements to establish a clear comparison between current ("as is") performance and potential ("could be") performance. This contrast provides tangible evidence of the financial benefits associated with solving quality-related problems.

For example, consider a situation where customer complaints are identified as a significant issue. A cross-functional team could be assembled to address the problem. The team might include representatives from sales (the team leader), customer service, manufacturing, engineering, and quality.

By applying COQ principles and structured problem-solving methods, the team can not only reduce complaints but also quantify savings, building a compelling case for both immediate corrective actions and long-term process improvements.

Step 2. Initiate Problem Solving

To be successful with structured-problem-solving COQ efforts, team members should have a basic understanding of problem-solving techniques. This doesn't mean they need to be experts, but a working knowledge of standard tools goes a long way. For example, team members should know how to write a clear problem statement, use a cause-and-effect diagram (also called an Ishikawa or fishbone diagram), and participate in brainstorming sessions to explore possible root causes. *The Quality Toolbox* published by Quality Press is an excellent resource for these and other tools.

In most organizations, it's the quality professional who recommends using COQ techniques for improvement projects. This person is usually already familiar with both COQ concepts and structured problem-solving. Still, if needed, the team leader or members of the implementation team can benefit from a refresher on the principles of quality costs. Even a short session can help spark new thinking and align the group around a shared approach.

In this example, the team examined patterns in customer complaints, focusing on complaint types and frequency. Complaint data is frequently available through the Customer Service or Quality Departments, which typically track complaints by category.

One particular issue—"no power-up"—stood out for its frequency and direct impact on the customer experience. The team agreed to make it the focus of their investigation.

As part of their analysis, they brainstormed possible causes and created a cause-and-effect diagram, illustrated in Figure 5.8.

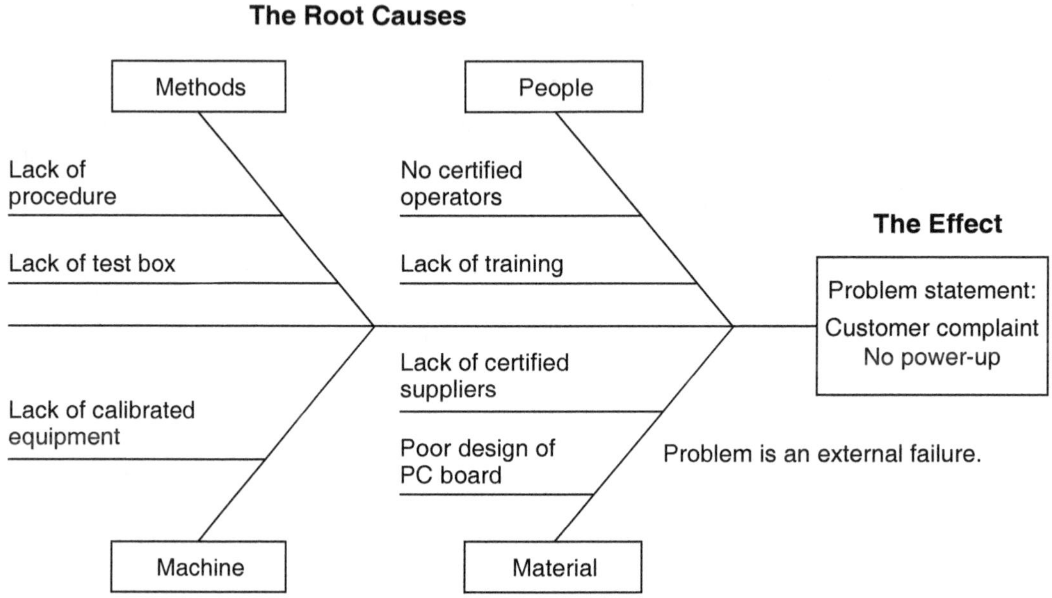

Figure 5.8 Cause-and-effect diagram.

Step 3. Calculate the Failure Cost of One Occurrence

Once a cause-and-effect diagram has been developed and the team has brainstormed potential root causes, the next step is to calculate the cost of a single occurrence of the selected problem. Using a COQ worksheet, similar to the one shown in Table 5.5, may prove valuable.

When building a COQ worksheet, keep it simple, especially at the start. Focus on the essential tasks required to identify and correct the issue. Begin by estimating the average time spent on each task and record time in decimal hours (e.g., *0.25* for fifteen minutes). Task time estimates can be derived from industrial engineering data or from asking those directly involved in the corrective action process.

For example, if six people work in the Customer Service Department and the total weekly payroll is \$8,350, the average hourly department rate is calculated as follows:

\$8,350 ÷ 40 hours/week = \$208.75/hour for the entire department
\$208.75 ÷ 6 employees = approximately \$34.80/hour per person

Table 5.5 COQ worksheet.

Tasks	Avg hours per task	Hourly rate	Cost of element	Material costs	Downstream consequences ($)	Waste ($)	Total failure cost
1. Customer Service answers the phone and records information	0.10	S35	$3.50	N/A	$3.50	N/A	$3.50
2. Forwards info to Quality Assurance (Q/A)	0.10	$35	$3.50	N/A	$3.50	N/A	$3.50
3. Q/A investigates the problem	1.50	$50	$75.00	N/A	$75.00	N/A	$75.00
a. Q/A performs destructive test on another stock part	0.50	$50	$25.00	N/A	$25.00	N/A	$25.00
b. Cost of part	N/A	N/A	N/A	$43.63	$43.63	N/A	$43.63
4. Fix the process in manufacturing	2.25	$50	$112.50	N/A	$112.50	N/A	$112.50
5. Contact the customer with fix and a new part	0.75	$50	$37.50	N/A $43.63	$37.50 $43.63	N/A N/A	$37.50 $43.63
Total							$344.26

In this case, the worksheet rounds this to $35/hour for simplicity. If multiple people are involved in a task, multiply the number of employees by the cost per task to estimate the total labor cost. Apply the same logic to tasks completed in other departments, adjusting for each department's average hourly rate.

The "Cost of element" column contains the product of time per task and the loaded hourly rate (including benefits). The "Material costs" column is used for actual out-of-pocket expenses—for example, replacement parts, inventory charges, unplanned travel, or any other direct costs incurred and paid.

When determining whether a cost should be categorized as waste or as a downstream consequence, identify where the defect was detected. If the customer finds the problem, the entire cost falls under downstream consequences. If it's caught internally, it's considered waste. Costs are not split between categories; each occurrence is assigned to one or the other.

Once all tasks and materials are listed, sum each row across to calculate the total cost per element; then sum the final column to determine the total failure cost. In the example provided, the total cost of one occurrence is $344.26.

To aid with tracking and trend analysis, it's helpful to assign a short code to each worksheet. In this case, the code *EF-NP* (external failure—no power-up) is written in the upper-right corner to reflect the problem type. This kind of shorthand can be helpful if data is later compiled into a reporting system or trend analysis dashboard.

It's important to note that these estimates do not need to be exact. COQ reporting is used to support management decision-making, not for formal accounting. As such, well-reasoned estimates are acceptable and often preferred for their speed and usability.

The "Hourly rate" column should reflect the average loaded rate (including wages and benefits) for each department involved. This information can typically be obtained from the Finance Department. If it's not readily available, a reasonable estimate can be made by dividing the department's total wages by the number of employees.

Step 4. Calculate the Total Failure Cost for a Given Period

Now that the cost of a single occurrence has been estimated, the next step is to determine the problem's frequency. This information can typically be gathered from existing sources, such as customer service call logs, defect reports, inspection sheets, or audit findings.

In our example, let's say there were forty-eight occurrences of the "no power-up" issue over the course of a year. With each occurrence costing $344.26, the total annual cost for this one issue would be:

$$48 \text{ occurrences} \times \$344.26 = \$16{,}524.48$$

A problem that may once have seemed minor or manageable now clearly warrants attention, especially when viewed through the lens of downstream consequences. By assigning real dollars to issues, teams can more effectively communicate their impact and urgency to management and other decision-makers.

Step 5. Rank Problems by Failure Dollars Using Pareto Analysis

The team can now repeat the costing process for other problems or defect types. By doing so, teams can evaluate both the frequency of occurrences and the quality costs, including waste and downstream consequences. The process becomes easier and more intuitive with each repetition.

In this example, the failure cost Pareto analysis worksheet, ranked by annual failure cost, would be organized as shown in Table 5.6.

Table 5.6 Pareto analysis of failure costs.

Problem description	Failure type	Annual failure $	No. of annual occurrences
No power-up	EF	$16,524.48	48
Computer downtime	IF	$11,838.11	20
The order shipped late	EF	$9,333.58	36
Paperwork incomplete	IF	$6,423.15	121

EF = external failure IF = internal failure

Notice that the most frequently occurring problem or defect does not always represent the highest failure cost. In this case, taking corrective action on the "no power-up" issue selected by the team will deliver the greatest return on investment. Quantitatively identifying the most significant contributor is where quality cost analysis proves its value, by helping teams prioritize problem-solving activities based on financial impact.

Of course, other factors, such as customer criticality, problem complexity, and the ease or timeliness of a solution, should also be considered. But before committing significant time or resources to address an issue, it's essential to have quality cost data that demonstrates a meaningful financial impact.

Step 6. Look at the Cause-and-Effect Diagram and Propose a Prevention Plan

It's now time to propose a plan to allocate investment resources to eliminate the root causes of the problem. At this stage, revisiting the cause-and-effect diagram shown in Figure 5.5 can help clarify which actions are most likely to prevent recurrence.

Continuing with our "no power-up" example, which carries an estimated downstream consequence cost of $16,524.48 annually, we can now calculate the cost of the proposed investment activities to eliminate the issue. These projected investments for prevention costs are outlined in Table 5.7.

Table 5.7 Proposed investment plan and costs.

Root cause description	Proposed tasks to eliminate	Hours needed	$ rate	Total $
Lack of procedure	Write procedure	3.0	$ 50/hr	$ 150
No certified operators and lack of training	Train and certify operators	40.0	$ 35/hr	$ 1,400
Lack of calibrated equipment	Calibrate equipment	2.0	$ 35/hr	$ 70
Lack of a test box	Make a test box	4.4	$ 50/hr	$ 220
Lack of certified suppliers	Find ISO-registered supplier	2.0	$ 35/hr	$ 70
Poor design of the PC board	None	N/A	N/A	N/A
Total investment required				$ 1,910

Step 7. Set Team Goal for Reducing Frequency of Occurrence

The next step is for the problem-solving team to estimate the level of reduction they believe is achievable if the proposed investment (prevention) actions are implemented. Depending on the issue's complexity and the strength of the identified cause-and-effect relationships, some teams may aim for a 100 percent reduction (a zero-defects goal). In contrast, others may set a more conservative target.

It's vital to avoid overcommitting. If unforeseen factors outside the team's control arise, an overly aggressive goal can be discouraging and may reduce engagement or momentum. A realistic, data-informed target is far more effective in maintaining focus and credibility.

In this example, the team set a 50 percent reduction goal for the "no power-up" issue, aiming to cut annual occurrences from forty-eight to twenty-four and, in doing so, eliminate half of the $16,524.48 in downstream consequence costs.

Step 8. Calculate the Return on Investment (ROI) and Payback Period

Once the reduction goal is set and the investment plan is developed, the next step is to calculate the ROI, also known as the benefit-to-cost ratio, and the payback period, which is the time required to recover the investment cost.

The calculation should be completed for each selected problem and organized in a worksheet like the one shown in Figure 5.9.

In our example, the external failure cost associated with forty-eight annual occurrences of the "no power-up" issue totaled \$16,524.48. An additional \$500 in unplanned premium freight charges, provided by the Finance Department, was added to this total, bringing the overall downstream consequence cost to \$17,024.48 annually.

The team committed to reducing the occurrence rate by 50 percent, resulting in expected annual cost savings of \$8,512.24. Potential savings are shown on the worksheet as the yearly reduction in failure cost.

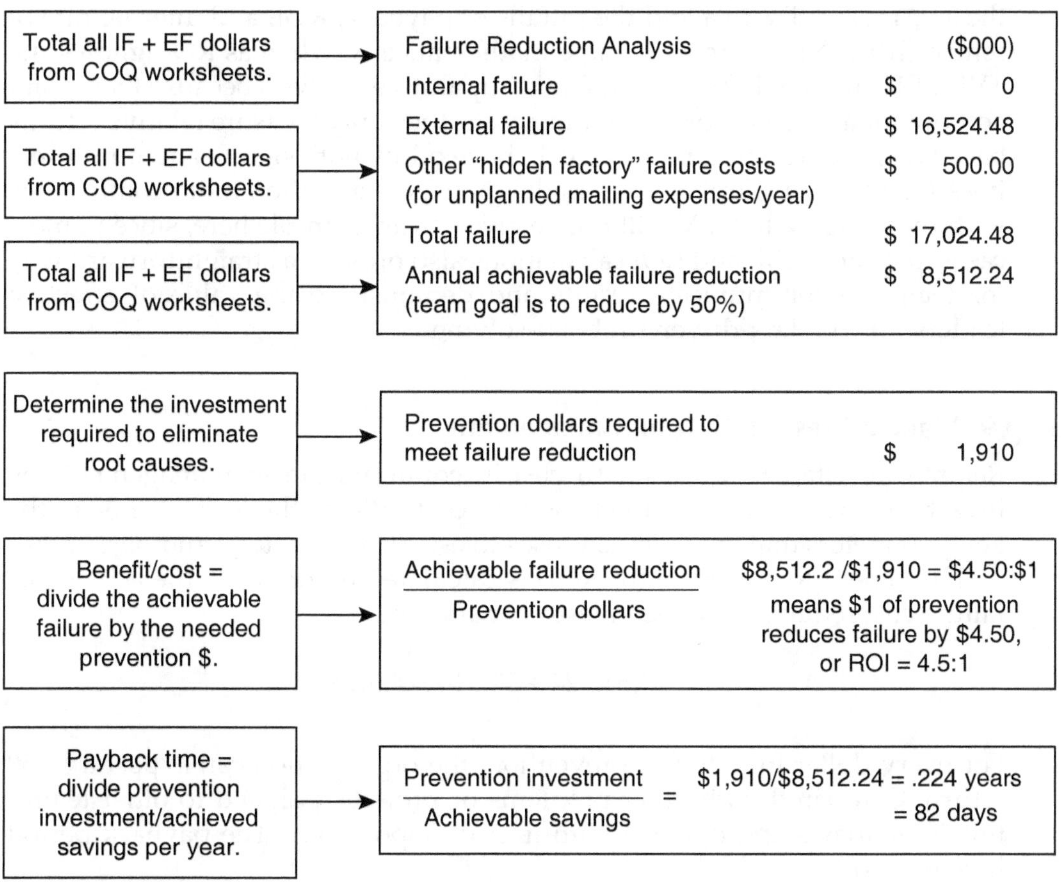

Figure 5.9 Return on investment and payback period worksheet.

Recall from the prevention plan above that the proposed investment was \$1,910. The benefit-to-cost ratio (or ROI) is calculated by dividing the achievable failure cost reduction by the investment amount:

$$\$8{,}512.24 \div \$1{,}910 = 4.5{:}1$$

This means that for every dollar invested in prevention, the organization can expect a $4.50 return.

The payback period is calculated by inverting the benefit-to-cost ratio:

$1,910 ÷ $8,512.24 = 0.224 years, or approximately 82 days

In other words, the $1,910 investment will pay for itself in under three months.

Both return on investment and payback period metrics can help teams and leadership prioritize issues and select the problems with the most significant financial impact and the fastest returns. It is worth noting that larger, more complex projects may necessitate more intricate calculations. To account for the TVM, use NPV to avoid the pitfalls of payback, ROI, and internal rate of return (IRR). NPV is most reliable mathematically, whereas ROI ignores the TVM. IRR and payback often fail in comparative analysis because they don't account for a project's total operating costs; they only measure relative returns based on proportions, and they can behave oddly with significant or multiple investments. The shortcomings aside, managers are often more comfortable with these. As such, NPV will not be given as an example here, since in most cases, payback, IRR, and ROI calculations also provide a straightforward way for teams to communicate results and demonstrate the value of effective leadership and data-driven problem-solving.

Step 9. Make a Presentation to Management

Referencing step 8, we need to clearly communicate to management the investment size and either the benefit-to-cost ratio or the ROI. These are the key parameters that management uses to establish a strategy and objectives.

The benefit-to-cost ratio (or ROI) is calculated by dividing the achievable failure cost reduction by the investment amount:

$8,512.24 ÷ $1,910 = 4.5:1

For every dollar invested in prevention, the organization can expect a $4.50 return. Compared with other problems or projects assigned to other teams, this issue may stand out as a high-impact opportunity. The payback period here is as follows:

$1,910 ÷ $8,512.24 = 0.224 years, or approximately 82 days

Emphasize that the $1,910 investment will pay for itself in under three months. Framing investment recovery will help leadership prioritize issues and select the problems with the most significant financial impact and the fastest returns. These calculations are a straightforward way to communicate results and demonstrate the value of effective, data-driven problem-solving to leadership.

Step 10. Track Progress and Repeat Process on Another Problem

As with any quality improvement project, progress must be measured and tracked. The technique presented here offers a straightforward approach: simply counting the number of occurrences of the problem. As occurrences decrease, the associated quality costs—particularly waste and downstream consequences—will also decline.

These numbers can be summarized by week, month, or any other appropriate reporting period, providing clear visibility into trends over time.

Once the prevention plan is implemented and measurable progress is achieved, the team can repeat this quality cost problem-solving process to address other issues and drive further improvement across the organization.

Using Six Sigma Where Variation Drives Cost

Since the inception of COQ as a formal discipline, the industry has experienced multiple waves of quality and performance management trends. Among such concepts as balanced scorecards (Chapter 10) and activity-based costing (ABC) (Appendix A) and such methodologies as reengineering and robust design, none has had a broader or longer-lasting influence than Six Sigma and lean. Of these, Six Sigma's rise to prominence brought with it an often understated but essential connection to the principles of COQ.

Before continuing, it's essential to clarify what Six Sigma is in relation to COQ. Six Sigma has evolved to mean a few different things in different contexts. While this distinction is not always formally acknowledged in the literature, it is generally understood in practice. One version of Six Sigma is a problem-solving methodology exemplified by the DMAIC (define, measure, analyze, improve, control) process and focused on reducing variation to near-zero levels (often cited as 3.4 defects per million opportunities). The other version is Six Sigma as a management system or organizational philosophy, best known through its large-scale deployment at GE under the leadership of Jack Welch. This approach involves formalized training structures (e.g., Green Belts and Black Belts), project selection frameworks, and executive reporting, all oriented toward enterprise-wide cost reduction and performance improvement. While both examples can and do often coexist, one functions as the management and distribution of the organization's resources, while the other is a structured problem-solving approach. Here, we are more interested in Six Sigma as a problem-solving methodology.

Six Sigma, as a management system, often borrows directly from the tools and logic of COQ, sometimes serving as a rebranding or restructuring of earlier quality cost practices. While its broad deployment has generated strong opinions both for and against, it's clear that, as a problem-solving methodology, Six Sigma has enduring value. It may not be the right fit for every situation, but it remains a practical and rigorous tool set for tackling complex, cross-functional problems rooted in uncontrolled variation.

COQ programs can benefit significantly from these tools once opportunities for improvement have been identified. When applied appropriately, the Six

Sigma framework provides a structured extension of the scientific method, enabling organizations to diagnose root causes and implement sustainable improvements rigorously. Some organizations may prefer PDCA over DMAIC. The authors suggest using whichever approach is most familiar and comfortable for the teams involved, provided the tools fit the problem and support the improvement goals.

Although Six Sigma as a management system has become less dominant in recent years, it can still be valuable, especially in organizations where experienced problem solvers are available but formal training programs and project infrastructure are less so. In either form, the methodology works best when the problem justifies the rigor: High-cost issues, cross-departmental inefficiencies, or performance gaps caused by variation are strong candidates for Six Sigma application.

In cases where the nature of the problem is different, more focused on speed, flow, or the elimination of waste, lean may be a better fit. While the two systems are often used in tandem, lean differs significantly in its philosophy and methodology, and it plays a unique role in quality cost management.

Lean Thinking Through a COQ Lens

Lean has had one of the most significant influences on organizations in recent decades. While many lean concepts are often seen as modern innovations, their core principle, delivering customer value without waste, builds directly on the legacy of early quality pioneers, both American and Japanese. Like Six Sigma and traditional total quality management (TQM), lean shares the goal of reducing variation and improving performance. Still, it does so with a scalable approach that can be effective on small segments of a process or on extensive processes, as in value-streaming exercises, and with a more intuitive focus on eliminating waste.

The concept of waste reduction is central to lean, and its straightforward language resonates with executive leadership. That accessibility and the practical clarity it brings are one reason the authors advocate using the term *waste* in place of *internal failure*. While the terminology shift wasn't originally intended to align with lean, it reflects the broader truth that the quality profession has, over decades, developed a deep understanding of waste in all its forms. That broader understanding of waste is what was intended, regardless of the category at hand or what categories are named.

Many practitioners, both inside and outside of traditional quality roles, are already familiar with lean's TIMWOODS framework, the eight forms of waste: transportation, inventory, motion, waiting, overproduction, overprocessing, defects, and skills underutilization. Others may recognize the equally viable alternative DOWNTIME. Lean provides not only the language but also the training and platform to identify and address these inefficiencies, making it a powerful complement to COQ systems.

When thoughtfully integrated, lean and COQ can become self-reinforcing. Much like the effective combination of lean principles and auditing techniques described in *Advanced Quality Auditing: An Auditor's Review of Risk Management,*

Lean Improvement and Data Analysis (Quality Press, 2015), lean and COQ programs can be harmonized to increase transparency, reduce inefficiencies, and improve outcomes. One powerful example is the use of value stream mapping, which puts customer value at the center of process analysis. By identifying non-value-added steps, organizations can highlight actual waste and justify quality investments earlier in the process, ideally during design and planning.

Other lean tools, like just-in-time (JIT), further enhance this alignment. Rooted in the principles of flow and pull, JIT is a direct response to the waste of excess inventory. While inventory carrying costs can be challenging to quantify, lean offers a simple directive: Eliminate waste where you can see it. In doing so, JIT can reduce not only inventory-related costs but also waste from internal defects and potential downstream consequences, such as customer complaints or returns.

Importantly, lean thinking can also be applied to the COQ system itself. By mapping the flow of information, resources, and decisions within COQ processes, organizations can identify inefficiencies in the collection, analysis, and utilization of quality data. These "second-order" applications of lean reveal how continuous improvement can occur not just on the production floor but at every level—from organization-wide initiatives to individual improvements at a workstation.

This fractal approach to improvement, where the same logic applies across scales, is one of lean's greatest strengths and a compelling reason to embed its principles in COQ programs. Whether you're targeting system-wide transformation or empowering individuals to own quality in their roles, lean offers a structured, accessible way to drive out waste, enhance value, and reduce total quality costs. Even if lean as a whole is not applied, the elements and broad scope of waste may be used in a general, structured problem-solving approach.

STUDY: USING TRENDS TO GUIDE DECISIONS

There are some commonly used techniques—such as trend analysis and Pareto analysis—applied to quality cost data. These can be used to break down costs by COQ category (investment, monitoring, waste, and downstream consequences); by element, department, product, service; or by any other meaningful grouping.

Trend analysis is one of the most straightforward and valuable tools in a quality cost program. At its core, it involves comparing current cost levels to past cost levels. Before drawing conclusions or planning improvement actions, it's essential to collect data over a reasonable period, typically several months, to establish a baseline and identify patterns.

This data should be plotted in multiple ways. Costs for each of the four primary COQ categories—investment, monitoring, waste, and downstream consequences—should be charted over time (e.g., monthly or quarterly). Results should be shown both as total dollars and as a proportion of relevant business metrics, such as sales revenue or units produced. This dual approach makes it easier to interpret results in context.

In addition, individual elements within each category—especially those contributing significantly to total cost—should be plotted and tracked separately. For example, Figure 5.10 shows total quality costs plotted over time, in total dollars, per unit, and as a percentage of sales. In that example, total costs are increasing, but relative costs (per unit and per sales dollar) remain stable, indicating that while costs are rising, they're not outpacing business growth.

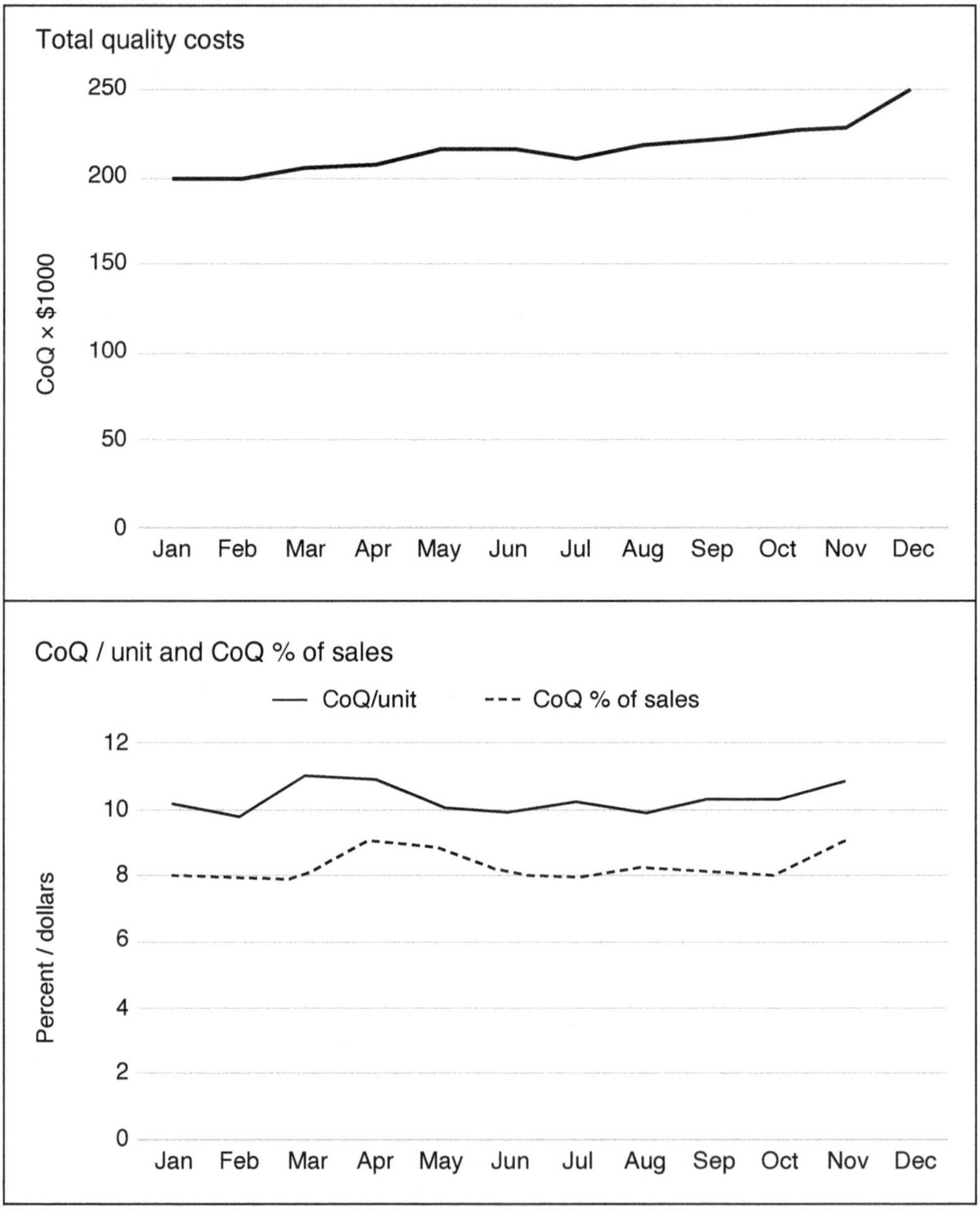

Figure 5.10 Total quality costs with/out volume adjustments.

Figures 5.11 and 5.12 go further by showing costs by category, again using both raw dollars and normalized metrics. In this case, all categories show rising total costs, but only waste shows a relative increase when measured against business activity. In this case, a preliminary analysis indicates the need for a deeper analysis, possibly using Pareto analysis, to identify the specific drivers behind the increase in waste-related costs.

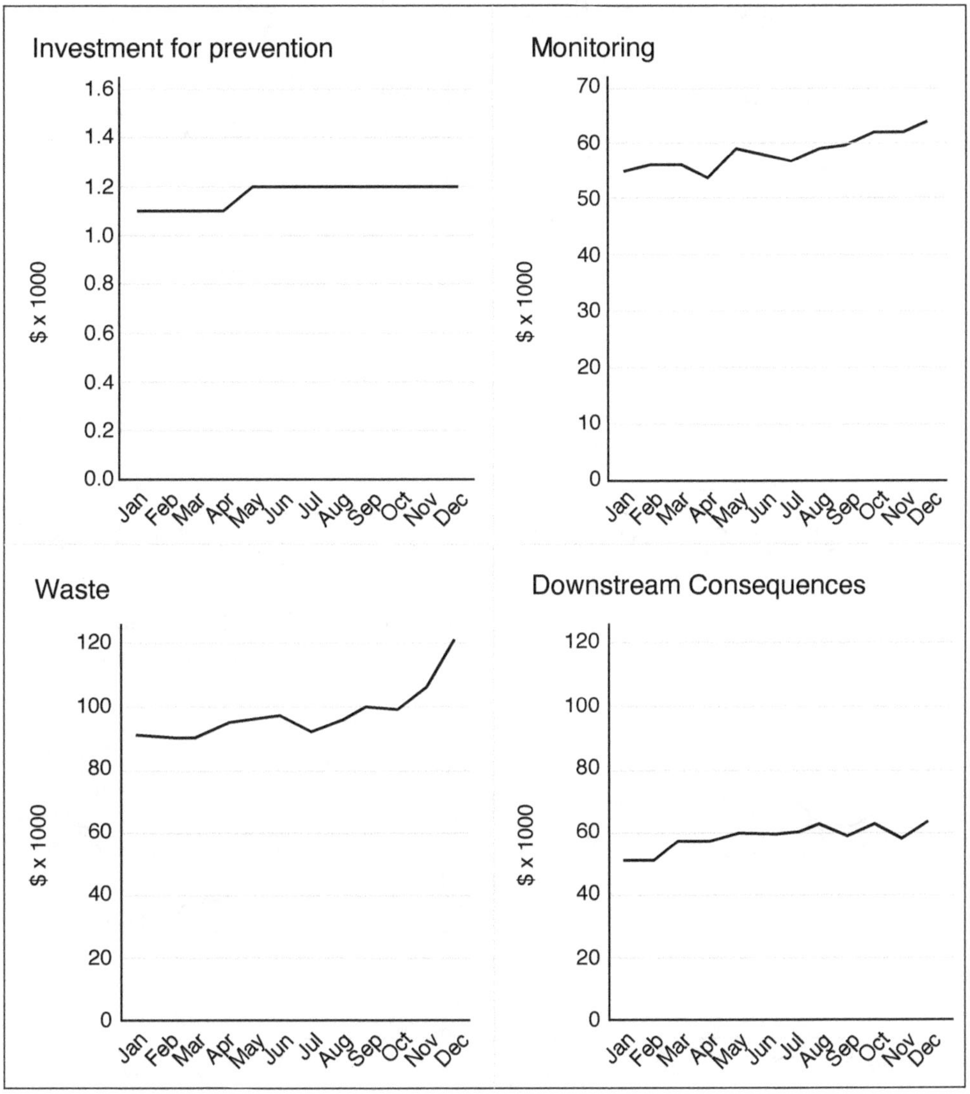

Figure 5.11 Quality costs by categories—total dollars.

There is, however, an important caveat when interpreting trends: Not all increases in quality costs are negative. As discussed in earlier chapters, purposeful reallocation of expenses across categories can indicate strategic improvement. In other words, it's often wise to invest more in investment and monitoring activities to intentionally reduce waste and downstream consequences.

This idea, often summarized as "an ounce of prevention is worth a pound of cure," is foundational to quality thinking. In most industries, it's far less expensive to catch and correct issues early (through investment and monitoring) than to deal with them later as internal rework or customer-facing failures. While especially true in regulated environments, such as medical devices or pharmaceuticals, where downstream consequences can carry significant legal risks, reputational risk is omnipresent; just consider any recent failure in the aviation or automotive industries.

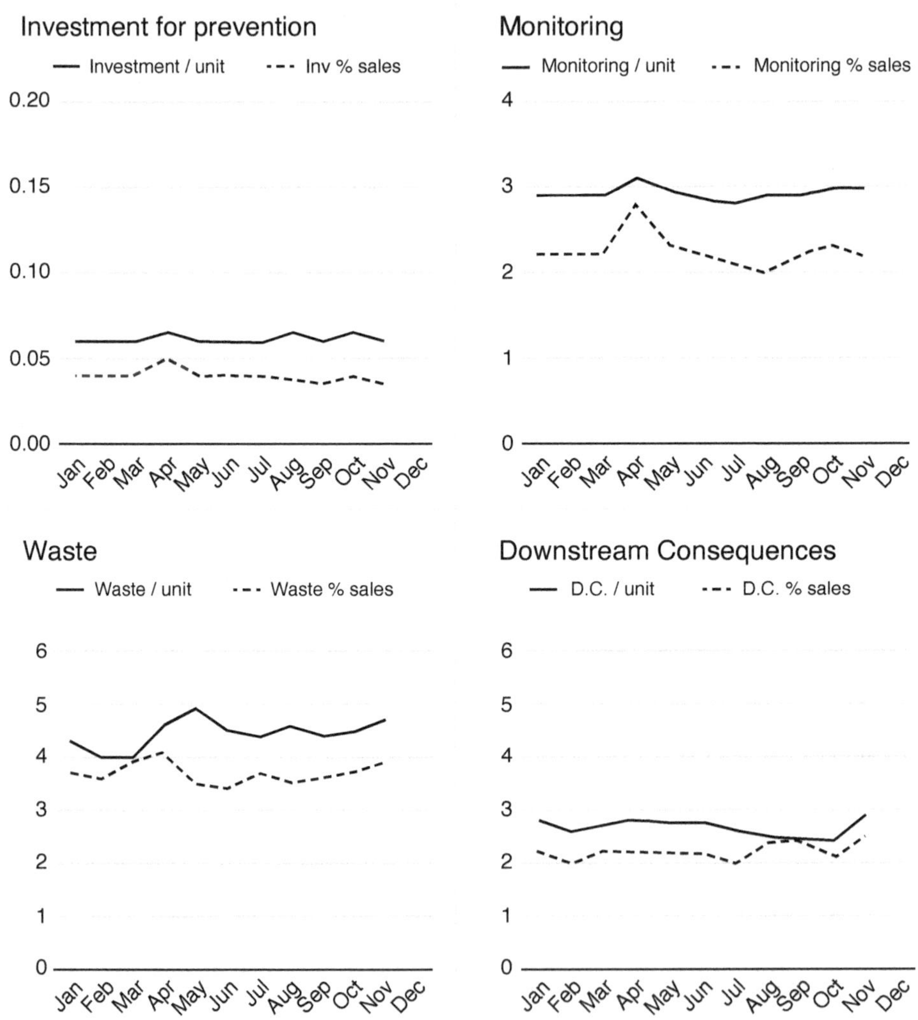

Figure 5.12 Quality costs by categories—related to bases.

While COQ trend data often shows a clear direction, it's essential to recognize the limitations of basic systems. Many initial implementations don't capture second- or third-order costs, such as return shipping, customer service labor, or the longer-term impact of lost future sales. Even so, the general principle remains widely accepted: When structured correctly, increased investment and monitoring often lead to reduced waste and downstream consequences—and, with maturity, even monitoring costs can be reduced through better planning and process control.

In other words, the long-term goal is not just to reduce failure costs but to build quality into the system from the start, through wise investment in design, training, and preventive planning.

This intuitive relationship between quality planning and quality performance is still often overlooked in rudimentary programs but should be embraced as part of a long-term strategy. And as discussed earlier in this chapter, your organization's evolution and structure are unique, so the way this relationship plays out will vary. The ROI calculation in Appendix A may help you justify and visualize these trade-offs when presenting proposals or building a business case.

Understanding and progress come from analysis and action that follow the numbers, but don't be afraid to invest a little to gain a lot.

ACT: INSIGHTS ON EXCELLENCE; APPLYING PATTERNS OF EXCELLENCE

This broader view of investment comes into sharper focus when we consider how companies allocate their resources. According to the ASQE/IOE 2025 survey, organizations aren't just expressing interest in poor quality costs—they're actively investing in quality with a variety of activities. Among respondents, 55 percent report funding quality training. Quality training typically reinforces the idea that building frontline capability remains one of the most practical levers for early error detection. Just shy of half, 43 percent, report applying structured methods, such as FMEA, lean, and poka-yoke, during operations planning, and a few more report investing focus upstream with customer research and robust design approaches like Design Failure Modes and Effects Analysis (DFMEA) and Six Sigma. We will cover some of this after the discussion of Pareto analysis.

What stands out is not only the diversity of these investment activities but their distribution across the entire value chain. These efforts aren't confined to a single department or phase. Instead, they stretch from early-stage marketing and design, the most reported after training at 46 percent, through supplier qualification, operational planning, to final validation and field feedback.

Among reporting organizations, 37 percent use customer surveys and focus groups to shape initial product or service design, while 44 percent invest in qualification testing, an area often overlooked but essential in avoiding issues before full-scale launch. Statistical process control tools were reported by 44 percent of participants. Interestingly, nothing is reaching universal or ubiquitous implementation, but the diversity of activities, with strong highlights, underscores the growing awareness that recurring quality issues frequently originate outside direct manufacturing.

This pattern illustrates a key point: Leading organizations view quality-related investment as a shared responsibility. It's not limited to a Quality Department or confined to a single phase in development. Instead, it's a coordinated strategy spanning design engineers, supplier managers, marketing analysts, and operations leaders. The goal is not to construct layers of inspection but to embed quality into the process itself through innovative design, capable systems, and proactive planning.

This mindset shift matters when justifying investment within a COQ framework. What might traditionally be labeled as overhead—training sessions, customer-insight work, or process-control tools—should instead be viewed as long-term cost avoidance. A well-timed investment in qualification testing, for example, may prevent an expensive product recall. Avoiding a warranty claim preserves not just financial margins but brand credibility and customer trust—intangibles that are difficult to recover once lost.

Organizations that balance their investments across the product life cycle often find it easier to implement and sustain quality cost systems. That's no coincidence. When cross-functional teams see how their work reduces internal failure and supports strategic goals, they're more likely to engage in cost tracking and continuous improvement. It shifts the conversation from reactive measurement to forward-looking decision-making.

So ask yourself: Are your current investments reducing future failure or simply addressing today's problems? Even small shifts in spending—toward design reviews, supplier readiness, or capability building—can yield disproportionately large returns. And when those investments are tracked against reductions in rework, delays, or support costs, the business case becomes far stronger.

In the long term, managing quality isn't about cutting costs wherever possible; it's about making targeted investments that eliminate the need for corrective action later. Investing early, across departments, and with purpose isn't just good practice. It's how high-performing organizations sustain results. This can be applied through the standard but indispensable Pareto analysis, as well as with such tools as lean, Six Sigma, or a structured problem-solving approach, depending on the problem and the team's knowledge and experience.

CONCLUSION

The methodology presented here offers a practical road map for calculating and acting on COQ data through team-based, structured problem-solving and analysis. Success or failure in implementing the COQ framework for improvement is not dependent on advanced statistics or complex tools. Instead, it relies on the relentless pursuit of simple, clear, and actionable fundamentals, which the COQ framework is designed to provide. It gives process improvement teams, managers, and other stakeholders a clear line of sight from everyday problems to measurable financial impact.

The example—showing a team leveraging COQ to guide an improvement project—highlights an often underappreciated truth: COQ is more than a ledger of costs. Used both as a management system and a problem-solving tool, COQ helps organizations surface and prioritize issues with real bottom-line consequences, including those that traditional defect-rate metrics (e.g., Parts per Million [PPM] or Defects per Million Opportunities [DPMO]) might miss. Applied this way, COQ enables teams not only to detect and control process changes or breakdowns but also to identify and prioritize improvements and secure leadership support to implement them.

KEY TAKEAWAYS

Regardless of the framework, methodology, or structure your organization or team selects, four guidelines are key. These four should be considered when implementing structured COQ problem-solving techniques or within your plant or enterprise-wide COQ initiative for improvement:

- *Focus on the problem and the process it triggers.* The structured COQ problem-solving method starts by identifying a specific issue and listing the tasks required each time it occurs. These tasks are then assessed using established COQ definitions, specifically, waste and downstream consequences. The result is a calculated cost per occurrence of the issue.
- *Integrate COQ into existing or planned teams.* There's no need to reinvent the wheel or label this as a "new and improved quality program." Instead, embed COQ into your existing teams' routine work. It can simply be incorporated into their weekly or monthly reporting process, using their existing meeting structures and tools.
- *Involve finance in reviewing calculations.* All COQ worksheets and underlying assumptions should be reviewed—and ideally approved—by finance. These techniques are not meant to replace or modify the formal accounting system. They offer logic-based estimates, not precise financials. Existing monitoring systems can often be leveraged to capture frequency and are readily available.
- *Provide training and support.* Some team members may struggle with COQ concepts or the structured problem-solving approach when first introduced. Offering basic training and clear guidance can ease adoption and build confidence across departments.

Chapter 6

Supplier Quality Costs—Managing Quality Costs Beyond Your Walls

All organizations depend on suppliers. Sometimes suppliers provide materials; sometimes they provide information; and sometimes, both. If your supply chain provides incorrect information or nonconforming materials, it results in quality costs for your organization.

Calculating the COQ within your supply chain is often more complicated than determining the internal COQ. The confusion arises from several sources.

Who is causing the issue, the buyer (i.e., your organization) or the supplier? Or is it a supplier's supplier?

The confusion persists when we consider who bears the cost of the issue. Who pays to prevent issues, who pays to monitor situations, and who pays to clean up the mess after the problem has been discovered? There may be contractual arrangements between buyers and suppliers under which the supplier must pay the buyer if an issue arises. Ensuring these topics are covered in your supplier agreements is standard contract management and part of the buyer's investment in prevention. The COQ goes beyond these formal payments. Many expenses related to supplier issues are only partially covered by these contracts. For example, a nonconforming shipment of material may be paid for or replaced by a supplier, but the problems it causes within the buyer's organization frequently involve hidden costs.

Another note about the supplier COQ is that the efforts a supplier makes to control their quality systems for a particular buyer add to their costs, which may be charged to the buyer. These charges may be identified as surcharges or hidden and consolidated into future quotes. It is an open question whether a buyer wishes to or even should involve themselves in the supplier's internal quality cost. Indeed, many suppliers are unwilling to share their internal quality costs with buyers to avoid giving buyers leverage in negotiations.

WHERE SUPPLIER PERFORMANCE HITS YOUR BOTTOM LINE

Supplier quality costs can be substantial—and when properly tracked, they serve as clear indicators of risk, inefficiency, and improvement opportunities within the supply chain. A system for managing supplier quality costs should follow the same general principles applied to other quality costs, using the four COQ categories defined in Chapter 2: investment, monitoring, waste, and downstream consequences.

Examples of some supplier-related quality costs include the following:

- investment costs, such as supplier quality audits, supplier quality surveys, or other qualification assessments;
- monitoring costs, such as receiving inspections or source inspections; and
- waste and downstream consequences costs, such as scrap, rework, rejected shipments, and site visits to address supplier-related quality issues.

Supplier quality is a strategic area where control is essential. While organizations often have limited authority to compel suppliers to make improvements, they do have influence—particularly when supported by clear, data-driven quality cost information. Control becomes the mechanism for detecting early signs of decline, holding suppliers accountable, and proactively managing long-term supplier relationships.

USING QUALITY COSTS TO MANAGE SUPPLIERS

If a buyer moves beyond simply requesting price reductions, quality cost data becomes a powerful tool for supplier management. The first step is to determine which supplier-related quality costs are high and measurable. Once categorized, a Pareto analysis can identify the "vital few" suppliers responsible for the majority of the cost.

From here, appropriate action can be taken:

- Encourage or assist key suppliers in implementing their own COQ systems.
- Collaborate on root cause analysis and process improvements.
- Reduce both visible and hidden costs through shared corrective efforts.

In some cases, requiring suppliers to implement COQ programs may not be practical, especially for small suppliers with limited resources. But even informal, data-informed improvement conversations can lead to significant gains.

Include clear, measurable quality metrics—such as defect rate, on-time delivery, and Quality Control Performance Index (QCPI)—in supplier contracts. Establish a regular review cadence (e.g., monthly or quarterly) to track these metrics and address trends early. Set up structured communication channels, such as scheduled performance calls or dashboards, to prevent surprises. When nonconformances occur, initiate a formal corrective and preventive action (CAPA) process with defined timelines and responsibilities for both parties.

USING COQ TO STRENGTHEN SUPPLIER SCORECARDS

One effective yet practical approach to integrating quality cost data into supplier management is to use it for a quantitative component of a supplier rating system.

A case study from an electrical products manufacturer illustrates this well. The company implemented a dual rating system: One part tracked traditional metrics (price and delivery), and the other tracked supplier quality and cost performance through a calculated index. We will not cover the price and delivery metric here, but we will concentrate on the quality cost index:

The formula for a supplier QCPI is as follows:

$$\text{QCPI} = (\text{supplier quality cost} + \text{purchased cost}) \, / \, \text{purchased cost}$$

While not every supplier cost could be tracked, the company focused on the most relevant and measurable expenses in its operations. Some administrative costs were not included (see hidden quality costs later in this chapter). A breakdown of example costs for Supplier C is shown here:

Cost of Processing Incoming Rejections

A study found that each rejected material lot required approximately $200 in documentation and return costs to the supplier. Therefore, the total cost of a supplier's rejected shipments over time was estimated as the number of rejected lots multiplied by $200.

$$2 \text{ rejected lots} \times \$200 \, / \, \text{rejected lot} = \$400$$

Cost of Complaint Investigations

A special study of the time needed to investigate complaints showed that this could not be estimated to any degree of accuracy. Therefore, each engineer was asked to document the time required for each supplier's investigation. The total cost of complaint investigations was estimated by multiplying the investigation time for that supplier by the average hourly wages and fringe benefits of an engineer.

$$10 \text{ hours investigation time} \times \$40 = \$400$$

Cost of Processing in Receiving Inspection

Because this company had labor standards for receiving inspections, this cost could be estimated by using the appropriate labor standard, the average hourly wages and fringe benefits of a receiving inspector, and the number of lots processed for a particular supplier.

$$1.00 \text{ std. hour} \, / \, \text{lot} \times \$20/\text{hour} \times 50 \text{ lots} = \$1000$$

Cost of a Defective Product After Receiving Inspection

After identifying an issue, it can be challenging to evaluate the impact because a defective product may be reworked or scrapped. If scrapped, the cost may not be recoverable from the supplier if responsibility cannot be clearly assigned. Additionally, sorting the remaining parts in the lot may be the best option. Fortunately, a study found that, regardless of the action taken, the purchase cost of that part provided an acceptable estimate of the quality costs incurred.

Therefore, this cost was estimated for Supplier C by multiplying the number of defective parts found after inspection by the part's initial purchase price.

100 rejected parts × \$2.80 purchase price / part = \$280

Caution: This method of estimating the cost of a defective product should not be adopted before verifying by a special study that it is reasonable under your specific circumstances.

This Supplier C quality cost equals the sum of four costs:

Processing incoming rejections = \$400
Complaint investigations = \$400
Processing in receiving inspection = \$1,000
Defective product altered receiving inspection = \$280
Supplier quality cost = \$2,080

The buyer also developed a method of interpreting the QCPI to assess other major suppliers. The QCPI was then calculated and used to classify suppliers based on their quality cost performance (Table 6.1).

Table 6.1 Five suppliers' ratings.

Supplier	Supplier quality cost (\$)	Purchased cost (\$)	Index (QCPI)
A	2,410	99,928	1.024
B	1,950	40,000	1.049
C	2,800	43,643	1.064
D	2,500	12,230	1.204
E	7,000	7,631	1.917

A perfect supplier would have no quality costs, since there would be no rejections or complaint investigations, and receiving inspection would be unnecessary. Therefore, the index for a perfect supplier would be as follows:

(Supplier quality cost + purchased cost) / purchased cost
QCPI = (\$0 + \$7,631) / \$7,631 =1.000

Table 6.2 Supplier rating scores interpreted.

Index (QCPI)	Interpretation
1.000–1.009	Excellent
1.010–1.039	Good
1.040–1.069	Fair
1.070–1.099	Poor
1.100+	Immediate corrective action is required

Immediate Corrective Action Required

Suppliers D and E in the example triggered the need for immediate corrective action based on this index.

Results for the overall program were encouraging: The percentage of total suppliers rated good or better increased from 75 to 80 percent, and supplier quality costs decreased by 8.5 percent in the first year.

Many variations and innovations can be developed for utilizing quality costs to evaluate supplier performance. How it is done is less important than recognizing the size and impact of supplier-related quality costs. Once identified, planned steps can be developed to measure, compare, and analyze, enabling improvement.

MOVING FROM CONCEPT TO ACTION

How to Begin a Supplier COQ Program

Provided an organization has progressed beyond simply asking suppliers for price reductions to keep the business, the buyer must determine which supplier-related quality costs to reduce. Comparing the relative magnitude of quality costs by category and element (see Appendix B for a list of elements) should be the first step. The buyer company's quality cost program could be an invaluable aid in accomplishing this analysis. For example, assume a situation in which purchased material rejections are the buyer's biggest problem. If the buyer believes quality costs will decrease as the material rejection rate improves, this is an essential issue for the company.

SUPPLIER COQ PARETO: FOCUSING EFFORT WHERE RETURN IS HIGHEST

Supplier-related quality problems are rarely distributed evenly across the supply base. A supplier COQ Pareto is used to identify where costs are concentrated, so management attention, supplier quality resources, and improvement efforts are applied where they will deliver the most significant value. The purpose is focus, not fault finding.

The analysis begins by defining a practical time window, typically a quarter, and a clear scope of supplier-related costs. Most organizations start by ranking suppliers, waste, and downstream consequences tied to supplier issues. Monitoring and investment can be added later if a fuller picture is needed, but they are not required for the analysis to be practical.

Costs are converted to dollars using data where available and standard cost assumptions where it is not. These estimates do not need accounting-level precision. The objective is correct ranking, not financial reporting. Directionally accurate numbers are sufficient if they consistently reflect the relative burden each supplier creates.

Supplier-related costs are then summed and sorted from highest to lowest. A cumulative percentage chart quickly shows whether costs are concentrated among a small number of suppliers. This ranking directly supports decisions about where to apply containment, corrective action, and supplier development resources first.

Before acting, the results should be verified. Large suppliers often appear at the top of the list simply because they ship more volume. For this reason, many organizations review both absolute cost and cost relative to purchased spend. The first shows where the money is going. The second shows where performance is weak relative to the dollars spent. Both views are important and address different management questions.

The Pareto should be rerun on a regular cadence. As costs decrease with one supplier, the next driver typically becomes visible. This keeps supplier quality effort aligned with actual financial impact rather than legacy priorities.

A Pareto analysis will likely show that a few suppliers are causing most of the problems. Now the buyer can focus effort on the "vital few" suppliers and take appropriate action. What is the appropriate action? The buyer might convince the vital few suppliers to implement quality cost programs. Discretion must be exercised before insisting on this. Some companies may be too small to support a quality cost program. Exceptional circumstances may exist in other companies that would prohibit this action. However, if a supplier determines that launching such a program is feasible, the quality costs visible to the buyer will most likely decrease. If these costs are reduced through effective CAPA and root cause analysis, the previously hidden costs borne by both buyers and suppliers will also be lowered. The result will be improved quality for both the supplier's product or service and the buyer's product or service. This should increase profits for both, and improved profitability for the supplier may eventually result in lower prices for the buyer if a strong relationship is established between them.

Engaging the Right Stakeholders

When developing a system to measure supplier quality costs, we have to begin with "who to involve" at the buyer's organization. The leadership in the procurement division needs to be involved. Their decisions, philosophy, and strategy will drive the supplier quality cost for an organization. In addition to procurement leadership, individual buyers within procurement need to be involved. Procurement specialists possess a wide range of skills and knowledge, particularly in understanding how suppliers react when issues arise.

In addition to the staff in the Procurement Department, product quality engineers will provide a quality perspective that may not be present among procurement buyers. Most procurement personnel focus on the purchase price or lead time. At the same time, they recognize the need to purchase high-quality materials, but they are less likely than a product quality engineer to identify potential shortcomings. Other staff involved in supplier quality cost measurement include the internal Cost Accounting Department and legal staff. Legal staff might seem an odd inclusion here, but most supplier-to-buyer relations are governed by contractual arrangements approved by the legal staff. When starting a supplier COQ measurement process, it is essential to review existing supplier contracts with the buying organization. These existing contracts may lack requirements to share information, or they may be structured to appear punitive toward suppliers.

It's essential to recognize that operations personnel often blame a supplier for substandard material when issues arise. This is an easy go-to solution, but it does not recognize shortcomings within the operating area itself. Placing blame on suppliers should be handled carefully, and most buyers acknowledge this. Recognize that suppliers may be crucial to helping an operation address internal problems. After all, the suppliers are likely to understand their materials better than the buyer. Also, suppliers may encounter issues with such material when shipping to other companies, potentially competitors of the buyer, and may be able to apply best practices learned from working with those competitors. Maintaining good communication and trust with suppliers is highly advisable.

Selecting Suppliers with Total Cost in Mind

It's essential to understand the need for effective supplier selection to weed out suppliers likely to incur significant quality costs for the buyer. Including a quality engineering perspective in surveys and site visits to assess individual suppliers' capabilities against buyer company requirements is a key consideration.

Larger procurement organizations may even deploy their own set of quality engineers to work with buyers. Whether the procurement organization creates a staff of quality engineers, or uses quality engineers from the rest of the organization while selecting suppliers, is immaterial.

Looking at the quality cost categories, expenditures for planning and supplier management generally fall under prevention investments. This would include planning for incoming inspection, source inspection, and tests necessary to determine the acceptance of supplier products. The cost of the tests themselves will fall under monitoring.

These quality-planning investments include preparing documents to ensure high-quality purchasing contracts. Also included in this investment is a review of the purchase order technical data, typically provided by personnel outside the Purchasing Department, to ensure the data clearly and entirely communicates accurate requirements to suppliers.

Monitoring That Prevents Escalation

Purchasing monitoring costs are allocated to ensure supplier parts, materials, or processes meet requirements and to minimize the impact of nonconformances. This includes inspections and tests, either at your site or the supplier's. Do these checks promptly to catch issues early.

Preventing Drift in Supplier Performance

Also, during monitoring, costs for additional inspections or tests are required to qualify the use of purchased materials. These costs are usually done once, but they may be repeated during multiyear production situations:

- first-article inspection (detailed inspection and worst-case tests) on a sample of the first production buy of new components, materials, or services;
- first-article inspection for second- and third-source previously qualified end-product key components;
- first-article inspection of the initial supply of customer-furnished parts or materials; and
- first-article inspection of the initially purchased quantity of goods for resale.

One common technique to control incoming materials is to establish a source inspection program. In this program, inspections are conducted at the supplier's site, typically by the buyer's personnel or a third party hired by the buyer. Of course, measuring the monitoring expenses for these programs must include all incurred costs, such as travel to the supplier's plant for monitoring activities and any independent test laboratory expenses.

Waste Due to Supplier Nonconformances

When a supplier provides nonconforming material, the buyer may incur numerous internal costs. These would include the cost of disposing of or sorting any incoming inspection rejects. It would be wise to include the documentation review and evaluation expenses resulting from the rejection, along with the disposition orders for handling and transportation. If the supplier is contractually obligated to pay for some of these expenses, these would not be counted as a COQ for the buyer. It is rare for 100 percent of these costs to be contractually obligated for the supplier to pay the buyer.

When material does not meet standards, it's not just the nonconforming material that needs to be dealt with; replacement material is also necessary. Therefore, the extra cost of replacing rejected materials should be included in the waste cost, whether they are discarded or simply returned to the supplier. Sometimes, replacement materials require significantly more expensive transportation due to tight schedules. The additional transportation costs for replacement materials should be added to the waste expense. Of course, if the supplier covers these costs, they would not be counted as waste costs for the buyer.

Due to time pressure to meet production deadlines, some buyers may undertake repairs on nonconforming materials that may not be billable to the supplier. These are part of the buyer's waste expenses.

The cost of performing corrective action is likely to be split between the supplier and the buyer. However, the expenses the buyer incurs to conduct failure investigations following supplier rejection are waste costs. If the material had been conforming in the first place, such analyses would be unnecessary. On rare occasions, a buyer may treat these investigations as an investment expense if the information gained is intended to enhance a robust quality control system.

What other actions can be taken if we know the magnitude of supplier quality costs? Incorporate these costs into your supplier rating system to improve accountability.

Costs You Don't See in the Purchase Order

While some supplier-related costs are easy to identify, many are hidden—just as with internal quality costs. These hidden costs generally fall into three categories:

1. Costs incurred by the supplier at their own facility
2. Costs incurred by the buyer while solving problems at the supplier's facility
3. Costs incurred at the buyer's facility due to supplier-related issues but not directly allocated to the supplier

1. Supplier-Incurred Costs (Hidden from Buyer)

Buyers typically lack visibility into internal quality costs at the supplier's facility. However, these costs are structurally similar to those the buyer incurs:

- investment efforts (e.g., product or process design, internal audits),
- monitoring activities (e.g., inspection by operators or quality staff),
- waste costs (e.g., scrap or rework before shipment), and
- downstream consequence costs (e.g., rejected parts or rework due to buyer complaints). This last category will be visible to the buyer as it occurs under their function.

For example, the supplier has made a significant investment. In a small machine shop, for example, expenses include product quality engineering. Someone must handle this task, and the production supervisor may well do so if the plant lacks a qualified engineering staff. The effort and cost still exist. Estimating small expenses may be done via interviews, audits, or surveys.

Certainly, effort is expended in monitoring, even by the smallest suppliers. Someone must inspect the product or service before delivery. (In a small shop, this is usually done by the operator who made the item.) Each shop, whether large or small, has waste costs. A supplier who makes a manufacturing mistake must either rework the item or scrap it, incurring a waste cost. If the supplier sends an imperfect item to the buyer, it may be rejected, creating a downstream consequence cost for the supplier. If the supplier is performing a service and the buyer is not satisfied, the supplier can either redo the work or replace the original product with a new one (e.g., a dry cleaner can replace a ruined garment), both of which are downstream consequence costs.

2. *Buyer-Incurred Costs at Supplier Sites*

These are costs the buyer assumes in solving problems on-site at the supplier's facility—such as travel expenses, quality engineer time, or crisis management efforts. These costs are rarely tracked or assigned to the supplier in most organizations, making them invisible in most accounting systems. A more robust buyer COQ measure will count these events and use typical activity cost estimates to produce a supplier-related waste metric.

3. *Hidden Costs Within the Buyer's Facility*

Supplier-related costs often go unnoticed within internal operations. Common examples include the following:

- designing custom gauges for receiving or in-process inspection,
- creating special specifications or controls to accommodate supplier variability,
- conducting additional inspection or testing on supplier material,
- calibrating extra equipment to assess supplier quality,
- lost production time or slowed production due to nonconforming material, marginal material, or late material, and
- field engineering or support to resolve supplier-induced product failures.

These hidden costs may be significant—and often represent missed opportunities for improvement or recovery. In medium- and large-sized organizations, accounting for such costs and allocating them to the procurement function when they are essential to the procurement function can drive better application of quality principles in purchasing, such as using a certified quality engineer with the buyer when vetting new suppliers.

These examples of supplier-related hidden costs do not exhaust all the possibilities. There are many more, some of which will be significant in individual situations.

ROI for Supplier Improvement

In the above example, Supplier E had a QCPI of 1.917, prompting discussions of contract termination. However, after a more careful assessment, the buyer determined that the supplier's systems and equipment were adequate and that the issue stemmed from start-up inefficiencies. The buyer applied ROI analysis to evaluate investments in supplier development.

First thoughts centered on canceling the contract with Supplier E and transferring the tools to either Supplier A, B, or C, all of which had much better quality performance indexes. A closer look at the situation revealed that Suppliers A, B, and C may not do much better initially, since this was a new product with significant start-up costs. A trip to Supplier E revealed that, although the company lacked a staff of problem solvers, the quality system and manufacturing equipment were adequate.

Rather than switching suppliers, the company sent a problem-solving engineer for two weeks to help stabilize operations. The decision was evaluated using a standard ROI formula:

ROI = (estimated savings ÷ investment) × 100
Estimated annual quality cost reduction: $6,000
Cost of engineer's time and travel: $1,500
ROI = (6,000 ÷ 1,500) × 100 = 400%

This proved to be a high-value intervention—and, in time, the supplier's performance improved while purchase costs declined. As an added benefit, supplier relations improved when the buyer helped the supplier implement changes that enhanced the supplier's operations for all its other customers.

Collaboration That Lowers Total Cost

There are always suppliers that stand out by creating difficulties. Some companies debit suppliers for scrap and rework occurring in the buyer's plant to shift responsibility for failure to the supplier, where it hurts most—in the pocketbook. In the long run, however, this may be counterproductive because some suppliers will ask for a price increase (or allow a price increase to creep in) to cover this situation. Another method often used by the buyer is to reduce business with the offending supplier, rewarding better suppliers with a larger "share of the pie."

While these actions are justified and acceptable, a more positive approach is to use supplier quality costs to identify needed supplier quality improvements. This approach can strengthen teamwork and trust between the supplier and the buyer. Better information exchange will help both firms reduce costs. The buying company can then initiate joint projects with suppliers to address the issues that drive high-quality costs.

The problems could be solved through buying company actions. The specifications may be incorrect, or the seller may not understand the component's application in the overall product. Ensuring correct specifications helps ensure the procurement of good parts and prevents them from being wrongly rejected.

In other cases, the supplier's manufacturing process may need upgrading through better tooling. By implementing joint projects that use supplier quality costs as a basis, these problems can be resolved, resulting in better products and lower costs for both parties.

A company should consider quality costs in its supplier relationships. This will allow the buying company to determine the costs and suppliers to focus on. The buying company can suggest to these suppliers that they adopt a quality cost program, if appropriate, or conduct special quality studies to improve product quality.

However, care must be used. Small companies may not be able to support this effort, and they may face exceptional circumstances that would prevent a successful application. Supplier quality costs can also serve as a basis for the buying company to initiate joint quality improvement projects with its suppliers.

Most important of all is that any quality costs program is incomplete without an effective corrective action program. The mere act of collecting quality costs will do nothing for your company but add expenses. Only by pinpointing and permanently solving problems can a company improve quality and productivity while reducing costs.

Final Thoughts on Supplier COQ

Supplier quality costs are often underreported, underestimated, and underutilized in supplier management. But once identified and structured through a COQ lens, they become a powerful basis for strategic supplier control.

Whether used for supplier ratings, ROI evaluations, or quality improvement planning, tracking and analyzing supplier-related quality costs provides both transparency and leverage. More importantly, they offer a pathway to collaborative improvement, enabling buyers and suppliers alike to reduce waste, improve performance, and create value for both parties.

The key is not just in measuring supplier quality costs—but in using the data to drive decisions, foster accountability, and prioritize long-term success.

KEY TAKEAWAYS

- *Supplier COQ is real money, not "supplier drama."* Nonconforming material and bad information create costs that often exceed what the purchase order shows.
- *Supplier COQ follows the same four buckets.* Treat supplier-related investment, monitoring, waste, and downstream consequences the same way you do internally.
- *The "vital few" suppliers usually drive most of the pain.* Use Pareto to focus effort on the highest-return areas.
- *Contracts help, but they do not cover the full cost.* Charge-backs and replacements rarely cover the costs of your internal disruption, delays, and rework.
- *Hidden costs are the difference maker.* Include buyer costs at the supplier site and the buyer's internal costs that never get assigned to the supplier.
- *Scorecards get stronger when cost is included.* Indexes like QCPI turn quality performance into a management number that leaders will use.
- *Supplier improvement should be evaluated like an investment.* ROI analysis can justify targeted support that beats switching suppliers, especially during start-up.
- *Collaboration reduces total costs more effectively than punishment.* The goal is fewer defects and less waste for both parties, backed by CAPA and shared data.

Chapter 7

Software Quality Costs

Economies are evolving, and just as services have grown faster than manufacturing, we now must acknowledge the rise of software. Software is significant in that it is both a product and a means of delivering products or services. Software delivers the most essential product—namely, information. Software captures and integrates information, linking customers to products and services and localities to global resources. Software also acts as a control mechanism for systems, transactions, individual projects, and enterprise-wide activities.

Software consists of computer programs, associated procedures, documentation, AI tools, and data about the operation of a computer system. As a purely intellectual product, it is among the most labor-intensive, complex, and error-prone technologies in human history. Differentiated from the hardware, the physical machines on which software works, there are many types of software products and systems in the world today. These are often characterized either by their business domains (e.g., information systems) or by their relationships to other software layered between the user and the computing machinery (e.g., programs embedded in microprocessors to create intelligent devices and appliances).

SOFTWARE IS DIFFERENT BUT NOT INCOMPREHENSIBLE

Understanding what makes software unique starts with comparing its development to traditional manufacturing. This contrast reveals why software projects behave differently—and why managing quality requires a different mindset.

- Software does not consist of physical components and therefore does not follow the Weibull "bathtub" curve. Software does not "wear out"; its obsolescence is reflected in the currency of its design and its interaction with other software.
- Software is predominantly a customized creation that is not replicated but somewhat enhanced or optimized with successive versions.

- Software is a non-material product rather than a physical product; its development is subject to human mistakes but not material flaws. It doesn't wear out.
- One cannot assume that a software specification is stable. Changing requirements is expected behavior in software development. (More discussion in iterative project management below.)
- Productivity levels vary widely (more so in individuals than in teams).
- Product defects result from insufficient or incomplete information, human misunderstandings, and human errors, not from deficient materials.
- Manufactured goods are valued for their features; software is valued for its features as well as its interactive functionality and nonfunctional attributes.
- The economics of software quality hinges on the process of understanding requirements.
- This process, more than conformance, is commonly responsible for a software product's value.
- The costs of the manufacturing phase of software production are insignificant. The bulk of software development costs is allocated to requirements elicitation and specification, design, implementation, and testing.
- Statistics cannot be applied to replications because software products are usually one of a kind.
- Software has significant security requirements and must support updates as hardware, operating systems, and other software change.

WHAT IS SOFTWARE QUALITY?

Although there is no single comprehensive standard definition of software quality, descriptions and terms are found in such sources as consensus national and international standards (e.g., ISO/IEC 25010:2011, ISO/IEC/IEEE 90003:2018) and other books on the subject (Figure 7.1).

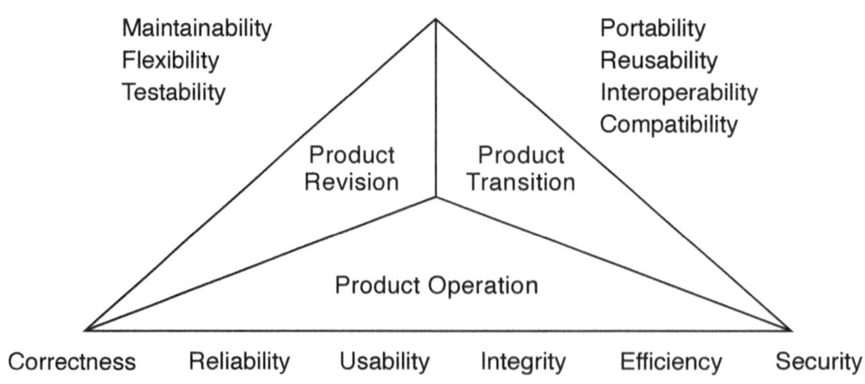

Figure 7.1 Software quality attributes.

ISO/IEC 25010:2011 is a standard that identifies and captures quality attributes for computer software. Replacing ISO/IEC 9126, it is part of a broader series of standards called SQuaRE (software product quality requirements and evaluation). It defines eight primary software quality attributes: functional suitability, performance efficiency, compatibility, usability, reliability, security, maintainability, and portability (Table 7.1).

Table 7.1 ISO/IEC 25010 software quality attributes.

Quality characteristic	Description
Functional suitability	The degree to which the software meets user needs, delivers correct results, and assists with desired objectives
Performance efficiency	The system's timely response, processing, and throughput are best achieved by using resources within limits.
Compatibility	The product can coexist successfully with other products and integrate with other systems.
Usability	Users see that the software meets their needs; they can learn it and operate it efficiently; user errors are minimized; the user interface is aesthetically appealing; and a wide range of users can use the product.
Reliability	The product has a low failure rate, is accessible when required, is tolerant of hardware or software failures, and allows restoration of data and operations after a failure.
Security	Data is accessible only to authorized users; unauthorized access to data and programs is prevented, events can be proven, actions can be traced, and user identity is verifiable.
Maintainability	Components can be changed with minimal impact on other components; components are reusable; impacts are analyzable; the system can be modified to improve, correct, or adapt; and the ease with which testing can be performed to determine whether requirements are met.
Portability	The software can be effectively and efficiently adapted to different hardware or software environments, including successful installation and uninstallation. It can serve as a replacement for another product for the same purpose.

Since software supports, enables, or automates professional services, it is reasonable to apply service quality models when evaluating the value and effectiveness of a software solution in a service context (Figure 7.2).

Service quality dimension	Examples	Relevance to software
Image, expectations	Appropriate to software use, alignment to the service model	Product "look and feel," replication of subject matter expertise; stakeholder and business requirements
Reliability	Complete transactions, fulfillment of commitments, success on first attempts	Confirmation numbers after transactions, traceability path, audit records, logs
Responsiveness	Prompt completion of steps, online or available assistance	Usability, online help, and frequently asked questions (FAQ) section
Regulatory requirements	Meets legal restrictions	International legal restrictions on privacy and security
Assurance	Instill confidence, make software users feel safe	Certification, security, disaster recovery
Empathy	Personalized customer care, retention of customer details	Customization, independent configuration, customer/ stakeholder communication management
Tangibles	Visually appealing, convenient	24-7 uptime and availability, works in different languages, and integration with other systems

Figure 7.2 Service quality model applied to software.

To summarize, it is crucial to reinforce the basic principles.

Here are some checkpoints to use to help guide software quality improvements:

- What is the degree to which a customer sees value in software products?
- How does the value in the software product compare to the competition?
- To what extent does the software possess a customer's desired combination of properties?
- How well does the software product perform for its intended use? Does it remain free from operational flaws?
- In the software development process, to what extent do the developers practically do the right things?

THE GROWING URGENCY OF MANAGING SOFTWARE QUALITY COSTS

If improving business success through software quality is a corporate goal (Figure 7.3), we need answers to a few simple questions that are rarely asked in today's software development environments:

- How much does software quality cost?
- What are the benefits of good-quality software?
- How good is our software quality?

Once the answers to these questions are known,

- quality costs can be compared to overall software production costs and software product sales;
- quality costs can be benchmarked against norms;
- deeper analysis can inform actions to improve the competitive position;
- the bottom-line effect of improvement actions can be measured; and
- the economic trade-offs involved with software quality become visible, thus leading to better decision-making.

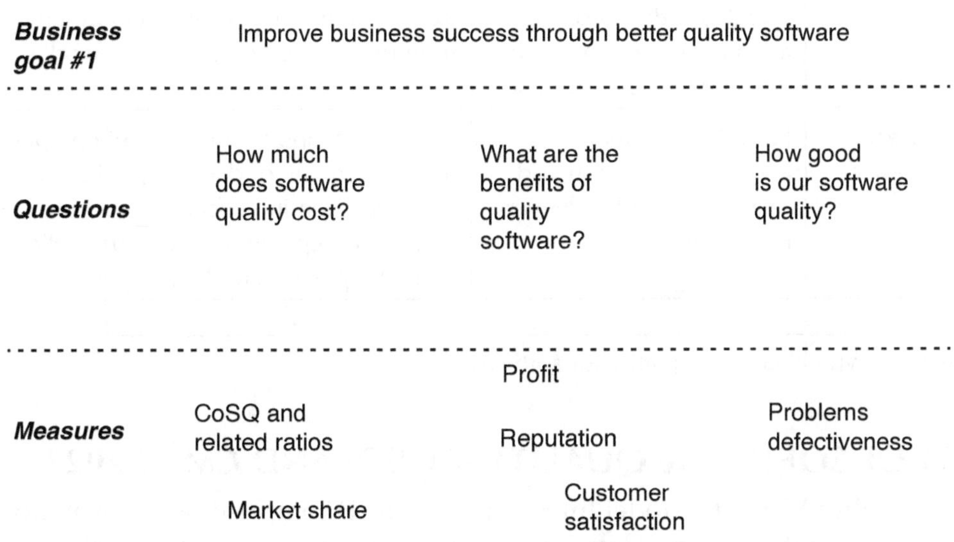

Figure 7.3 Cost of software quality in context.

APPLYING COQ PRINCIPLES TO SOFTWARE

In software project management, quality costs fall into four categories (Figure 7.4). Understanding these helps you decide where to invest and where to cut losses.

Category	Description	Typical costs
Investment	Defect avoidance, quality basis definition, project- and process-oriented interventions	Efforts to define quality and set quality goals, standards, and thresholds; quality trade-off analysis; definition of release criteria for acceptance testing and related quality standards; training; process metric creation and planning; testing environment, beta testing
Monitoring	Defect detection, discovery of product nonconformance, and finding the level of nonconformance	Quality control gating processes, contract or proposal reviews, quality audits, go/no-go decisions, quality assurance of subcontractors, inspections, static/dynamic analysis, user acceptance testing, walk-through, desk checking
Waste	Prerelease defect or anomaly correction before delivery to the customer	Recode, retest, rereview, redocument, requirements rework, design rework
Downstream consequences	Postrelease defect or anomaly correction and related costs after delivery to the customer	Warranty support, resolution of complaints, reimbursement for damage paid to the customer, domino effect on other projects, damage to reputation, added marketing to correct reputation problems, penalties

Figure 7.4 Typical costs of quality for software.

COST OF SOFTWARE QUALITY (COSQ) AND CMMI 2023

Capability Maturity Model Integration (CMMI) comprises a series of models: development, services, data, managing security and safety capability, and planning and managing work capability. Users of the CMMI Institute–supported models (CMMI standards are now owned by ISACA, a global professional organization focused on IT governance, cybersecurity, risk management, and assurance) select predefined sets of model components applicable to different business environments.

Much like applying COQ in a custom way for each organization's context, environment, and strategy, each organization selects its capability area, then its applicable practice areas, then practice groups, and finally chosen practices. The high-level capability areas are grouped into the categories of doing, managing, enabling, and improving. Looking deeper under doing, the capability areas are

- ensuring quality,
- engineering and development products,
- delivering and managing services, and
- selecting and managing suppliers.

The practice areas under ensuring quality are

- peer reviews,
- process quality assurance,
- requirements development and management, and
- verification and validation.

Each of these practice areas is organized by capability levels 1 through 5, although the possible number of maturity levels varies across practice areas.

It is at this practice-area level that the COQ applications begin. Of the four practice areas under the ensuring quality capability area listed above, all are likely to be involved in our monitoring CoSQ, except requirements development and management, which is more likely to be in the investment CoSQ.

WHAT ORGANIZATIONS GAIN FROM MANAGING COSQ

Implementation of these practices has led to demonstrable organizational improvements across many performance measures, including schedule and cost predictability, product and service quality, productivity, customer satisfaction, and ROI.

One of the benefits of measuring and using CoSQ is justification for quality initiatives. Examples of ROI in quality improvement initiatives speak most clearly to managers responsible for maintaining a profitable organization. However, CoSQ can be used for several other benefits. It can

- provide cost data for motivational purposes by demonstrating the relationship of employee efforts to the bottom line,
- provide a basis for budgeting the quality operation,
- compare and identify the most cost-effective process improvements,
- provide a measure for comparing the success of projects, and
- identify quality improvement candidates.

GETTING STARTED WITH COSQ IN SOFTWARE

Consider the business case for the CoSQ program. Why are you doing this? How does it support the overall strategy? A small pilot program in a limited area for a limited time may work to illustrate the value of CoSQ.

There are many methods to collect costs for a CoSQ program. Most organizations have a cost accounting chart of accounts, and some of these accounts may be directly and wholly linked to elements of CoSQ: investment, monitoring, waste, and downstream consequences.

For more details, some quality costs can be taken directly from individual departmental accounting reports. In other cases, it may be necessary to rely on basic accounting and engineering records, such as schedules, time reports, defect reports, and purchasing records. In worst cases, one may resort to working with members of the software organization to estimate each quality cost element.

The exact sources and methods will vary with organizational size and sophistication, as well as whether traditional waterfall project management is used, some form of interactive project management (such as *agile*) is used, or a combination of both.

COMMON CHALLENGES IN SOFTWARE COSQ PROGRAMS

One of the pitfalls of a CoSQ program concerns *controversial cost elements.* Typically, the question concerns which costs are normal operating costs and which are quality costs. An example would be the cost of producing a project management plan. Although this plan is intended to manage a project's expenses and schedule, it also influences product and process quality.

It is helpful to keep in mind the following points:

- The quality of the data used needs to be confirmed. Inaccurate or erroneous data will hurt overall credibility.
- Arguments over controversial elements have been known to undermine program quality.
- The most significant quality costs are those that are most readily discernible (reviews, software quality assurance, testing, and rework). Therefore, it is often safe to exclude controversial elements without unduly affecting the total CoSQ.
- Consistency throughout a CoSQ program is more important than thorough inclusion of quality costs, because it enables clear identification of improvements and potential improvement candidates.
- It is essential to begin somewhere and to adjust the cost allocation as experience is gained.

Concerns may also arise as to how quality costs should be categorized. Again, consistency is important. For example, the costs associated with very early peer reviews may be treated as investment expenses rather than as monitoring costs. This is a matter of interpretation, depending on when a work product is considered ready for review. In software production, inspections may be incorporated into the production process in accordance with the peer review CMMI capability area ensuring quality. For documentation purposes, an item is not considered complete until it has undergone peer review and revision. The same is true for code, especially when code inspections precede unit testing; clearly, it is a monitoring activity.

MANAGING COSQ ACROSS PROJECTS AND PORTFOLIOS

In organizations that complete much of their work through projects (true of most software firms), there is a three-way responsibility break:

- Projects are temporary; they create a unique result or product, satisfy a distinct objective, and deliver business value to the organization.
- Programs are related activities that may include multiple projects and are directed as a coordinated set of efforts.
- Portfolios are collections of projects and programs driven to achieve a specific business objective. There may be significant dependencies across projects and programs, and portfolio management serves as a layer that optimizes resources to support the overall business strategy.

In this breakdown of responsibilities, the four CoSQ categories will vary across functional levels. See Figure 7.5 for a possible breakout of CoSQ for projects, programs, and portfolios.

CoSQ Category	Project	Program	Portfolio
Investment	Quality training and education	Quality training and education	Quality planning
		Supplier qualification	Supplier qualification
		Supplier quality planning	Supplier quality planning
Monitoring	Process capability evaluations	Quality system audits	Quality system audits
	Testing of new or enhanced software	Analysis review	Analysis review
	Test beds	Test beds	Process, product, and service audits
	Qualification of the supplier's products, including software tools	Qualification of the supplier's products, including software tools	
	Verification and validation activities	Process, product, and service audits	
Waste	Completing failure reports and tracking them to resolution	Investigating failures to identify defects	Software that was created but never used

Figure 7.5 CoSQ in projects, programs, and portfolios.

CoSQ Category	Project	Program	Portfolio
	Rebuilding the software to include corrections, verification, and validation activities performed on the correction		
Downstream consequences	Recording user failure reports and tracking resolution	Recording user failure reports and tracking resolution	Warranties, performance penalties, and litigation
	Correcting user-identified defects	Product recalls	Product recalls
	Technical support services, including help desks and field service	Technical support services, including help desks and field service	Loss of reputation or goodwill, customer dissatisfaction and lost sales

Figure 7.5 CoSQ in projects, programs, and portfolios *(continued)*.

MANAGING COSQ IN ITERATIVE AND AGILE ENVIRONMENTS

In iterative project management, there will be multiple sources for CoSQ. The *scrum* framework, for example, uses evolutionary, adaptive, and self-correcting practices to address some of the challenges of traditional waterfall project management. An example of a challenge with conventional waterfall planning is extensive planning of activities that prove unhelpful when demands shift (e.g., customer requirements or unexpected difficulties), causing projects to fall quickly behind schedule and over budget. Iterative project management techniques share a simple idea. When you're working on a project, regularly check in to see if what you're doing is heading in the right direction and if it's still aligned with stakeholders' expectations. Along the way, look at what you're doing, decide if there's a better way to do it, and what might be keeping you from making those improvements.

Fundamentally, this entails applying multiple plan-do-check-act cycles within small teams. Performed by close-knit teams within relatively short time frames, this activity ensures that, upon completion of each subrequirement, the subsequent requirements are coherent. Each one of these cycles is called a *sprint*, and a daily stand-up meeting is used to guide the work. These standups have a few rules:

- Meetings are held at the same time each day.
- Everyone on the team needs to be present.
- The meeting can't last more than fifteen minutes.
- Everyone has to participate and be focused on moving things forward.

Tracking Quality Costs in Agile Teams

It stands to reason that CoSQ will differ when interactive project management techniques are employed. Recall that in the traditional waterfall project management method, the overall project is planned at the beginning (investment) and then executed. While being executed, the plan's accuracy is measured (monitoring), and corrections (waste) tend to accumulate toward the end of the designated project time frame. Examples of CoSQ for traditional projects are shown in Figure 7.6, to be used to add to (or replace) the elements found in Appendix B.

With its rapid cycling and small, incremental steps by smaller teams, iterative project management breaks the investment in planning into smaller increments that occur daily or weekly. Monitoring is necessarily more adaptive, and as the team progresses, there may be changing needs for measuring activities. Waste in iterative project management is reduced but may persist; occasional rework is required. The downstream consequences may be similar to those in waterfall project management, in that once the project is complete, field complaints must be addressed.

Another aspect of iterative project management requires a higher-level view. If multiple small teams are working simultaneously on numerous requirements for a software package, coordination among them is needed. Occasionally, one team will perform a task that affects another team's activities; this must be communicated to the affected team promptly. Such rework continues to be called waste, but this shared communication between teams might be called monitoring, not at the project level.

Some organizations are working to become nimbler and embrace the idea of "failing fast." This means pilot projects or small investments and an acceptance of the risk of failure. This is a cultural shift at organizations that has affected CoSQ. The background is that it is better to fail early and incur lower waste costs than to face higher downstream consequences or continued investment costs. From this idea, we can see a significant difference between traditional waterfall project management and iterative project management.

Given the inherent differences in iterative project management, the traditional categories in Figure 7.6 must be adapted for an iterative organization. It is important to remember that in fast-moving iterative projects with small teams, the burden of overmeasurement can significantly retard progress. A balance must be struck between the team's quick thinking and the organization's broader need to improve processes.

CoSQ category	ID	Description
1 Investments	1.1	Requirements
	1.1.1	Marketing research for customer/user quality needs
	1.1.2	Customer/user quality surveys
	1.1.3	Product quality risk analysis

Figure 7.6 Example of traditional CoSQ elements.

CoSQ category	ID	Description
	1.1.4	Prototyping for customer review
	1.1.5	User requirements/specification reviews/inspections
	1.2	Project
	1.2.1	Project quality planning
	1.2.2	Project process validation and verification
	1.2.3	Quality assessment of development platform and tools
	1.2.4	Platform and tools development for quality
	1.2.5	Developer quality training
	1.2.6	Quality metrics data collection
	1.2.7	Design for quality: software component reuse
	1.2.8	Formal inspections/peer reviews
	1.2.9	Project configuration management
	1.2.10	Project change management
	1.2.11	Supplier capability assessment
	1.3	Reuse library
	1.3.1	Salaries
	1.3.2	Expenses
	1.3.3	Training
	1.3.4	Platform and tools
	1.4	Configuration management administration
	1.4.1	Salaries
	1.4.2	Expenses
	1.4.3	Training
	1.4.4	Platform and tools
	1.5	Software Quality Assurance administration
	1.5.1	SQA Salaries
	1.5.2	SQA Expenses
	1.5.3	Software process and standards definition and publication
	1.5.4	Metrology: data maintenance, analysis, and reporting
	1.5.5	SQA program planning

Figure 7.6 Example of traditional CoSQ elements *(continued)*.

CoSQ category	ID	Description
	1.5.6	SQA performance reporting
	1.5.7	SQA education/training
	1.5.8	Process improvement, including team meetings and activities
	1.5.9	SQA process compliance audits
2 Monitoring	2.1	Supplied product testing
	2.2	Project appraisal costs
	2.2.1	Verification and validation activities
	2.2.2	Testing: planning, platforms, setup, test data generation, test execution, and logging, reporting, test data evaluation, test beds, test databases, test tools
	2.2.3	Product quality, process quality, and service audits
	2.3	External appraisals
	2.3.1	Process maturity evaluation
	2.3.2	Field performance trials
	2.3.3	Special product evaluations
3 Waste	3.1	Product design defect costs
	3.1.1	Causal analysis and reporting
	3.1.2	Design corrective action
	3.2.3	Rework and retest due to design corrective action
	3.1.4	Work products were wasted due to design changes
	3.2	Purchased product defect cost
	3.2.1	Defect analysis cost
	3.2.2	Cost of obtaining the product fix
	3.2.3	Cost of defect workarounds
	3.2.4	Rework
	3.3	Implementation defect costs
	3.3.1	Defect measurement and reporting
	3.3.2	Defect fixing
	3.3.3	Causal analysis and reporting
	3.3.4	Project process corrective action

Figure 7.6 Example of traditional CoSQ elements *(continued)*.

CoSQ category	ID	Description
	3.3.5	Fix inspection
	3.3.6	Retest and integration
4 Downstream Consequences	4.1	Technical support for responding to defect complaints element
	4.2	Product returned due to a defect
	4.3	Maintenance and release due to defects
	4.4	Defect notification costs
	4.5	Upgrade due to a defect
	4.6	Service agreement claims (warranty expense reports)
	4.7	Litigation costs and liability claims (insurance and legal reports)
	4.8	Penalties (product contract reports)
	4.9	Costs to maintain customer/user goodwill due to dissatisfaction (sales reports)
	4.10	Lost sales/market share due to quality problems (field salesperson reports)

Figure 7.6 Example of traditional CoSQ elements *(continued)*.

Sources of quality cost data: Ordinarily, quality cost data for most categories would be obtained from salary and expense reports. Exceptions in the downstream consequences category are shown here in parentheses.

INTERPRETING COSQ BEHAVIOR OVER TIME

Plotting CoSQ against a quality measure (such as total cost of ownership or product defects) yields the CoSQ curve and, when plotted over time, reveals trends in an organization's quality processes. This addresses most of the goals of quality costing: justification, motivation, budgeting, and process improvement cost effectiveness. Pareto analysis, based on the principle that quality costs are localized (80 percent of quality costs are incurred by 20 percent of the quality nonconformances), can be used to identify candidates for process improvement. When CoSQ is categorized by product and process sources, typically one or two sources will be shown to incur much higher costs than the others.

When analyzing CoSQ data, the model's dynamics must be considered. The following are some of the factors that affect CoSQ over time:

- Advances in software technology, for example, the prevalence of graphical user interfaces, create new demands on software producers.

- Growing user sophistication creates demand for increased functionality and better support (e.g., documentation, training, technical services).
- Better applications and systems set new standards for performance and reliability.
- Organizational digital transformation will increase training needs.
- Better software development technology, for example, in testing and configuration management, supports higher-quality goals.
- The application of increasingly powerful AI technologies is affecting development time, the handling of potential AI glitches, and human oversight.
- Reduced cycle time and time to market (accelerated change cycles) forces more planning in a limited timeframe.
- Changes in actual or perceived quality may accompany increased competition in the marketplace.

These factors may shift the point of diminishing returns over time, usually to the right. For example, either improved software development technologies or increasing user sophistication can shift the optimal CoSQ toward higher quality.

MAKING COSQ METRICS MEANINGFUL

For presentation to leadership, relationships that have the most significant impact are as follows:

- quality costs relative to the total cost of ownership for software,
- quality costs as a percent of sales,
- quality costs compared to profit, and
- quality costs compared to the magnitude of the current problem.

In addition to these, showing CoSQ as a percent of total development costs is appropriate for software for several reasons. First, sales and profit may not have a direct relationship with the actual cost of a software product, since market forces often determine software pricing. Second, almost all software development costs can be measured in labor hours, allowing them to be readily expressed in either hours or dollars. Third, the state of the art in software development is such that comparing quality costs to development costs illustrates the magnitude of the current challenge.

A COQ TOOL FOR AGILE

We can break quality costs into four categories: prevention investment, monitoring, waste (internal rework), and downstream consequences (external failure). To manage the CoSQ in agile environments, we need to shift from *unplanned rework* (waste) to *risk mitigation* (investment). The core philosophy is to identify a "sweet spot" in which the total cost of these costs is minimized rather than aiming for zero waste cost.

A Sprint Quality Dashboard helps to visualize the tension between feature velocity and quality costs, using a 1:10:100 Rule to illustrate how early investment exponentially reduces long-term costs:

- $1 spent on investment (requirements/design).
- $10 spent on monitoring (testing/QA during the sprint).
- $100 spent on waste (fixing it during other sprints).
- $1,000 spent on downstream consequences (fixing it after the customer finds it).

Some Examples of CoSQ Management Software for Agile

The software tools shown by the specific COQ category help measure or manage.

Investment and Monitoring Tools

- SonarQube/SonarCloud: The "gold standard" for tracking technical debt; provides time/dollar estimates for fixing code smells and vulnerabilities.
- Selenium/Cypress: Used to quantify monitoring costs by tracking time spent writing and maintaining automated test suites.
- Snyk: Identifies security vulnerabilities early to assess the investment cost of security.

Waste and Process Tracking Tools

- Jira: Used to calculate the rework ratio (bugs vs. stories) and track cycle time to identify time lost when tickets move backward in the workflow.
- Linear: A streamlined alternative to Jira that offers built-in triage and cycle time metrics to measure waste (defect) management.

Engineering Intelligence Platforms

- Jellyfish/LinearB: These sit on top of Git (a version control system [VCS] in agile development that enables teams to track code changes and manage different software versions) and Jira to provide investment distribution reports, showing the percentage of capacity spent on features, maintenance, or waste.
- Downstream Consequences (Error Tracking and Agile Project Management [APM] focusing on the process of delivering software)
- Sentry/LogRocket: Tracks real-time production errors to quantify the cost of hot fixes performed outside the normal sprint cycle.

- Datadog: Tracks service downtime, allowing teams to calculate the COQ based on average revenue lost per hour.

To explain this to leadership, clarify that the team is currently paying a "stability tax" on every feature. Use a stability ROI calculator, like the one in Table 7.2.

Table 7.2 Stability ROI calculator.

Investment item (the "fix")	Investment expense (current sprint)	Risk of waste (if ignored)	Estimated downstream consequences (postrelease)	Net savings (ROI)
Refactor legacy auth module	20 hours ($2,000)	40% chance of 80hr outage	$8,000 + reputational loss	$1,200 (expected value*)
Automate regression suite	40 hours ($4,000)	100% chance of 15hrs/ sprint manual QA	$1,500 per sprint (ongoing)	Payback in 3 sprints
Fix "flaky" CI/CD pipeline	10 hours ($1,000)	2 hours lost per dev/week (5 devs)	$1,000 per week	$3,000 saved/ month
Address critical security debt	15 hours ($1,500)	5% chance of data breach	$100,000 + (legal/ compliance)	$3,500 (risk adjusted)

Notes:

1. Choose three to five items that are currently slowing the team down.
2. This example assumes an average fully loaded engineering cost of $100/hour.
3. Explain like this: "Because our build process is flaky, every feature we ship takes 20 percent longer than it should. Over a quarter, that costs us $X in lost productivity."
4. Leaders love knowing when an investment starts paying for itself. Present the "break-even point" like this: "If we take one sprint now to automate our deployment, we break even in six weeks. From that point on, we are effectively 'earning' ten extra engineering hours every week for free."

CONCLUSION

Software is a special case in the global economy and in COQ, yet the lessons reach far beyond code. Because software is both a product and a delivery system, it underscores that quality must be aligned with the requirements, design, and workflow. Its shift from the waterfall model, of project management, to agile demonstrates how prevention, monitoring, and control can be embedded in rapid cycles without sacrificing rigor. The same principles apply to any product or service.

What matters most for progress is how COQ connects. Implementation begins with clear definitions and practical measures that finance and operations can trust. Control is about consistency, cadence, and ownership, so that gains endure and rework does not recur. Improvement is about identifying the significant drivers, eliminating causes, and reinvesting savings in prevention.

Seen this way, software is not an exception but a mirror. The same COQ logic that brings discipline to code brings clarity to factories, hospitals, and service networks. COQ is both the language of money and the logic of quality. If you let it guide how you implement, control, and improve, it becomes more than just a cost measure. It becomes a means of deciding, sustaining, and creating value across the enterprise.

KEY TAKEAWAYS

- *Software is a special kind of "product."* It does not wear out, it changes, and its value comes as much from interaction and nonfunctional performance as from features.
- *Most software quality cost is upstream.* Requirements and design work drive the economics far more than "production" because replication is cheap and misunderstanding is not.
- *The four COQ buckets still work.* Investment, monitoring, waste, and downstream consequences map cleanly to software when you define examples that teams recognize.
- *Use standards to keep quality concrete.* ISO/IEC 25010 provides a practical set of attributes leaders and teams can point to when "quality" gets fuzzy.
- *CoSQ answers three management questions.* What does quality cost? How good are we? And what do we get back for investing in quality?
- *Start practical and avoid the "category wars."* Exclude controversial elements if needed but be consistent so trends and improvements are believable.

Chapter 8

Case Studies

Quality costs originated in manufacturing, but their relevance extends far beyond the factory floor. Every process, in an office, a clinic, a software team, or a university, contains unproductive cost. When defects slip through, customers notice, and the organization incurs costs in lost time, rework, and reputational damage. The logic is simple: Downstream failures cost far more than upstream prevention. Investing in robust process controls and targeted improvements not only reduces waste but also often yields returns at ratios as high as 8:1.

This chapter explores how COQ principles apply across diverse environments. Each case study demonstrates how organizations identified waste, measured its impact, and acted to improve performance. You'll see patterns in four key categories—investment, monitoring, waste, and downstream consequences—and learn how decisions changed because of the numbers. For practitioners, the goal is practical: Use these examples as templates, not absolutes.

Ask three questions as you read:

1. What was measured, and how?
2. Which decisions shifted because of the data?
3. How were gains sustained through control plans and review cadence?

The cases that follow—from office workflows to healthcare projects, software development, and beyond—show that COQ is more than a metric. It's a management tool that connects quality to financial outcomes, enabling leaders to move from reactive fixes to proactive, predictive improvement.

CASE 1: COQ IN THE OFFICE ENVIRONMENT

Reference

Wood, Douglas C. "Types of Waste in the Office and Their Relative Magnitudes." Master's thesis, University of Central Missouri, 2014.

Over the last several decades, employment in the United States has shifted from primarily manufacturing based to, for want of a better word, office based. The majority of workers in the US currently work in office environments. These offices produce numerous outputs across numerous industries, but the work environment is relatively similar from one industry to the next. This similarity allows us to view the office COQ as a collective system.

Office work is not precisely the same as service businesses. Office work can provide services, but it can also be the managing and coordinating force behind manufacturing. Indeed, office work is used across industries from healthcare to construction. In many ways, it doesn't matter what the specific outputs are because the nature of the tasks and the work from an office are so similar.

Since internal failure cost in traditional COQ categories is really waste, we can turn to another area that discusses waste: lean. We know from studying lean manufacturing that there are seven classic wastes: transportation, inventory, motion, waiting, overproduction, overprocessing, and defects. A cursory examination of these seven classic wastes indicates they don't fit as directly in an office environment.

Categories of Office Waste

In "Types of Waste in the Office and Their Relative Magnitudes" (2014), various approaches to classifying office waste were explored. In that thesis, the following categories were identified based on interviews and surveys: meetings, distractions, email and other messaging, information storage and retrieval, health issues, motivation, poor results, and workflow. Subsequently to the publication of the thesis in 2014, another category emerged: waiting on systems or software.

The research used a survey that allowed individuals to self-assess the extent of time lost due to these multiple categories of office waste. The anonymous survey asked applicants to indicate their income level voluntarily. This allowed the cost to be calculated once the survey results were collated. Although the final category of waiting on systems was not included in the survey, the thesis estimated, based on self-identified waste, the dollar value of these categories of office waste, expressed in terms of costs per person per year in Table 8.1 with definitions in Table 8.2.

Table 8.1 Data from 2014 survey for annual cost of office waste in 2025 dollars.

Category	Median annual cost of waste per person (2025 dollars)
Workflow	$6,820
Distractions	$4,940
Messaging	$4,740
Motivation	$4,063
Meetings	$2,248
Information delays	$2,162
Health issues	$697
Total	$25,670

Table 8.2 Definitions of office waste categories.

Category	Definition
Workflow	Decisions and/or work involving wasted time
Distractions	Activities not related to regular/assigned work
Messaging	Time wasted in emails, phone calls, voicemail, and texting
Motivation	Missing personal motivation due to various reasons
Meetings	Work gatherings for no useful purpose
Information delays	Searching for information
Health issues	Issues that interfere with work
System or software delays	Waiting for software responses

Highest Cost Categories

In the decade since the survey was conducted, it's worth considering changes in our office environments. Workflow is identified as the most significant source of office waste. Survey respondents indicated that much of this workflow waste involved redoing work because the first pass did not meet requirements. Office automation has increased dramatically over the past decade, although not all office work has been automated. This automation may reduce flows, but if it is not appropriately aligned with requirements, it can also increase rework.

The next largest category of waste is distractions. In the survey, respondents were asked to identify how much time they lose to distractions in the office environment and to estimate how often those distractions provide additional benefits, for example, in problem-solving. This was done because not all distractions are harmful: Sometimes, being distracted from a task can lead an individual to identify a solution to another problem. In the survey, time spent on negative distractions is subtracted from time spent on positive distractions. It remains the second-largest category of office waste.

With more work-from-home employees, these distractions may have been reduced. However, this would be unlikely if individuals working from home are not properly isolated from home-based distractions. Expanding this thought to automation, automated activities in offices don't require human oversight, but humans are still involved in those transactions, even if the core of them is automated. The distractions of human participants in automation still cost time and, therefore, money.

The third-largest category of office waste is messaging (unnecessary communication, such as emails to extraneous recipients). It is doubtful that messaging waste has been meaningfully reduced in the last decade. There are more ways than ever to send messages from one office worker to another. There is still no consensus on the best methods for using our technologies to message most effectively. Messages still get lost, whether they are email, text messages, voicemail, or other forms. This category may have become larger as face-to-face messaging declines.

There have been many articles written about the effect of low morale and poor motivation in today's office environment. It is beyond the scope of this book to characterize the costs of motivation issues, but they are real.

The category of meeting waste is still significant on the above list. Anecdotally, everyone has been in meetings that were held for no good purpose. The techniques for improving meeting effectiveness have been around for many decades, yet they are not universally applied.

Measuring these types of wastes can be challenging, but it's not impossible. To measure the cost of waste in an office, you can use a survey such as was used in the thesis. You can also look for metrics that can serve as proxies for some of these wastes. Several techniques are available to determine the time an office worker spends without contributing value.

CASE 2: COQ IN HEALTHCARE

References

Wood, Douglas C. "Applying Lean Principles to Clinic Operations, Part 1." *Nephrology News and Issues* 25, no. 12 (2011): 24–27.

Wood, Douglas C. "Applying Lean Principles to Clinic Operations, Part 2." *Nephrology News and Issues* 25, no. 13 (2011): 20–22.

Many waste costs in healthcare are universal. While hospitals have made significant progress in applying quality techniques and practices (notwithstanding the results), most hospital cost accounting systems lack provisions for tracking the cost of internal waste. The cost of waste in healthcare situations can be measured on a project-by-project basis.

In a series of project-based COQ measurements involving a series of previously published lean healthcare projects, of the seven projects studied, only one agreed to publication upon completion of the COQ estimates. The reasons for these refusals to publish are not officially stated. Still, these seven studies showed significant financial shortcomings in the management of these hospitals; unsurprisingly, leadership in these institutions did not wish to have the cost of their mistakes publicly attributed to their institutions.

Even without specifics, the case of the seven is interesting, and we can examine further the one study that allowed a COQ publication: a lean application in a kidney dialysis clinic in eastern Europe.* The lean project was very successful in streamlining operations at the clinic, reducing cycle times, and boosting staff morale. The COQ results would not be expected to differ if it were a US healthcare facility, as kidney dialysis is the same in both countries.

The investment for this project is presented in Table 8.3.

Table 8.3 Investment costs for Regional Center of Nephrology in Poland.

Investment activity	Performed by	Annual cost
Analyze assets and usage	Quality engineer	$96
Staff discussions on methods	Quality engineer, nurse, clinic physician, cleaning staff, ward manager of nurses	$916
Brainstorming session with team	Quality engineer, nurse, cleaning staff, ward manager of nurses	$142
Design and print equipment markers		$3,685
Design flowchart for main processes	Quality engineer, CEO	$758
Wet floor sign		$89
Floor tape (red yellow)		$356
Document organization	Quality engineer	$203
Implement different uniform color for staff	CEO	$94
Visual Management (VM) implementation	Quality engineer	$128
Kanban implementation (+ kanban cards)	Quality engineer, CEO	$594
Total investment cost		$7,061

The monitoring expenses are summarized in Table 8.4. Only the listed monitoring costs were incurred after the change; however, other costs may have been incurred before the lean project. 5S is a lean manufacturing methodology designed to improve workplace efficiency, safety, and organization by adopting five Japanese principles: Seiri (Sort), Seiton (Set in order), Seiso (Shine), Seiketsu (Standardize), and Shitsuke (Sustain).

Table 8.4 Monitoring costs for Regional Center of Nephrology in Poland.

Monitoring activity	Performed by	After annual cost
Process owner reviews 5S every day (10 min)	Cleaning staff	$955
Visual board update every week (15 min)	Quality engineer	$416
5S audits (30 min)	Quality engineer, cleaning staff	$284
Meetings with quality team in order to audit actual procedures (including kanban system)	Quality engineer, nurse, cleaning staff, ward manager of nurses, clinic physician	$916
Total monitoring cost		$2,571

The increased monitoring costs were modest. The waste identified in the lean project is listed in Table 8.5.

Table 8.5 Internal waste costs before and after for Regional Center of Nephrology in Poland.

Internal waste activity	Performed by	Before annual cost	After annual cost	% improved
Nurses searching for items and preparing orders	Nurses at dialysis station	$34,488	$10,346	70.0%
Wrong use of equipment or items (e.g., yellow Wet Floor signs, misplaced equipment)	Nurses, cleaning staff, CEO	$680	$20	97.0%
Excess inventory (kanban cards, more stock than needed)	One-time savings	$53,400	$ -	100.0%
Drugs out of date (because of disorder, few locations)	Two warehouses instead of 4; more accurate ordering	$28,480	$712	97.5%
Cleaning equipment costs	Monthly purchases	$8,544	$6,408	25.0%
Breakage of equipment (caused by wrong layout, no VM, not including general technical service)	Unusual technical service	$2,848	$ -	100.0%
Ordering logistics costs by various employees	Ward manager/ CEO known material locations	$2,242	$560	75.0%
Total internal waste cost		$130,682	$18,046	

Waste before the lean effort and waste present after the project are listed, along with the percentage improvement resulting from the project. Most lean projects do not measure the costs of investment, monitoring, and waste before and after implementation, as they focus on operational improvements, such as cycle times.

A summary of the results and payback for the lean project is in Table 8.6.

Table 8.6 Results and payback for the lean project.

Category	Before annual cost	Investment	After annual cost	Net savings	Savings to investment ratio
Investment		$7,061			
Monitoring			$2,571		
Internal waste	$130,682		$18,046	$112,636	
Total	$130,682	$7,061	$20,617	$110,065	15.6:1

A 15:1 return on investment is not unusual. Some of the lean projects that were not eligible for publication had ratios as high as 700:1.

Since no comprehensive COQ data is available from healthcare institutions, and this study (along with the others not published) indicates a significant project-by-project benefit, we may conclude that there is much good work to be done with improving healthcare awareness by applying quality cost principles and encouraging leadership to pursue improvement projects, no matter how they are named.

* Katarzyna, R., C. Michał, J. Paweł, and M. Justyna. "Polish Dialysis Center Employees Use Visual Management to Enhance Safety and Improve the Organization of the Medical Facility." *Administrative Science Quarterly* 69, no. 1 (2024): 1–28.

CASE 3: PROJECT MANAGEMENT APPLICATION

Reference

Wood, Douglas C. "Project Oriented COQ." *Quality Management Forum* (2008).

The Project Environment

While quality costs have been applied in a wide variety of firms, project-based work possesses several characteristics that make quality costs difficult to apply:

- Projects have unique scopes.
- Projects operate in unique environments.
- Project priorities often change during the progress of the project.
- Projects can't tolerate yield loss. Projects must be complete to be acceptable.

First, each project has a unique scope. While many projects may be similar, each successive project will require careful understanding of costs and timelines during planning and execution. From project to project, the scope, timing, size, and resources will vary, and an effective project manager must adapt their approach to each project.

Second, each project takes place in an altered environment. Outside forces are likely to vary significantly across projects, making comparisons between projects more difficult. Different tasks may occur in each new project, so the quality cost measures in each project can differ qualitatively and quantitatively. While projects are still likely to fall into groups (in the context of quality costs), these groups may not be identifiable using the usual parameters.

Third, as a project starts, the project managers will have different priorities than at its close. The quality costs incurred at the beginning will primarily be investment or monitoring costs, while waste costs will accumulate at the end of a project. This eliminates the need for a steady-state analysis of quality costs, an approach used in mass-production environments. At different points during a project, the ideal quality cost proportions of investment, monitoring, and waste will look different.

Fourth, a mass-production system can tolerate a degree of failure. Every mass-produced item has a tolerable level of poor quality. While this level varies across products, it can be relatively high. In a project-based business, every project must meet its goals. A single failed project can devastate a project-based company, regardless of its size. In some cases, a larger enterprise can suffer more from a failed project because of its greater visibility. A project manager needs a guide to understand how a quality cost measure will affect the outcomes of their project.

These characteristics operate synergistically to place quality costs in a different light for project-based businesses compared to mass-production companies. Such significantly different demands suggest a new model may be needed, one that better aligns the practice of quality costs with project work.

Another Model of Quality Costs for Project-Based Environments

We imagine a time chart of quality costs for a project. The chart begins at week 1, with an initial investment value for a start-up. Over time, investment costs peak and then decline. Monitoring costs increase as the project progresses, then peak and decline toward the end. Waste costs begin to accumulate as the project nears its end, peaking just as it winds down. See Figure 8.1.

Although each project has a different chart, there are likely similarities across them. A chart would need to be adjusted for project length and cost to eliminate scope differences. The remaining more complex differences will be due to fundamental characteristics of the project and its environment.

Comparing project differences can identify the optimal level of investment and monitoring to control failure costs. Projects can be classified by their quality cost characteristics, and future projects can be planned to achieve better outcomes.

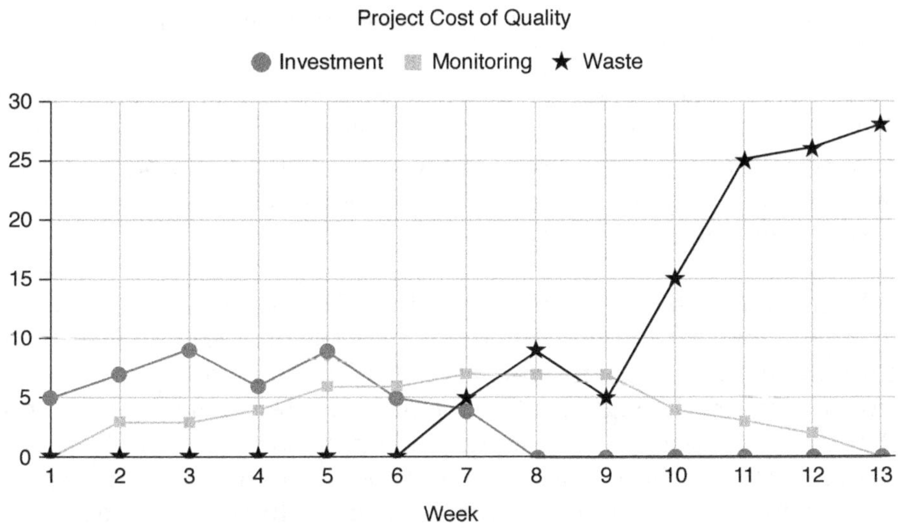

Figure 8.1 Project-based COQ over time.

Standard quality cost models, built around multiple project cost histories, help guide projects, providing a project manager with indicators for the best levels of monitoring and investment activities.

Need for Real-World Data

There are several basic questions here:

- How does project scale affect the investment-monitoring-waste data curves?
- Are there different patterns for classes of projects?
- What are the key drivers for different levels of investment and monitoring costs?

Several prominent engineering firms have adopted this approach using time-and-attendance data collected from their staff. The time-and-attendance data is coded to separate the COQ categories, and the shape of the curves characterizes projects. Standard curves have been developed for specific project types. This data is not generally available because each engineering firm considers its approach a source of competitive advantage and is unwilling to disclose how it handles it publicly.

CASE 4: SOFTWARE APPLICATION AT RAYTHEON

References

Haley, Thomas J., Raylene Swanson, Neil Rogers, and Edward R. Baker. "Software Process Improvement at Raytheon." *IEEE Software* 12, no. 6 (1995): 33–41. https://doi.org/10.1109/52.469758.

Dion, R. "Quantifying the Benefit of Software Process Improvement." Paper presented at the SEI/AIAA Software Process Improvement Workshop, Chantilly, VA, November 8, 1990.

Krasner, Herb. "Accumulating the Body of Evidence for the Payoff of Software Process Improvement." In *Software Process Improvement*, edited by Robin B. Hunter and Richard H. Thayer. IEEE Computer Society, 2001.

COQ has long been a proven tool in both manufacturing and service industries—for communicating the value of quality initiatives and for identifying opportunities for improvement. The same principles apply to the software industry through CoSQ, though its use remains relatively limited.

Early applications of CoSQ suggest that poor software quality can account for a substantial portion of development costs—often exceeding 60 percent in organizations with limited awareness of improvement opportunities. However, the potential for savings is significant. For example, Raytheon reported a fourfold reduction in software rework costs after implementing targeted quality improvement initiatives informed by CoSQ analysis.

Beyond cost reduction, CoSQ provides critical insights into the economic trade-offs associated with quality-related decisions in software development. By quantifying these costs, organizations can better evaluate investments in testing, defect prevention, process improvements, and other quality-enhancing activities—ultimately enabling more informed, data-driven decisions throughout the software life cycle.

Introduction

This case study provides visibility into a benchmark software development organization that currently uses CoSQ to document the results of improvement programs and to observe how CoSQ concepts relate to the consistent production of high-quality software. The Raytheon Electronic Systems (RES) organization was selected as a case study because, using a CoSQ approach, it demonstrated specific, measurable benefits from its software improvement efforts. Moreover, they collected data and lessons learned about their CoSQ implementation activities.

Although RES is in the contract systems business, it is similar in many respects to a commercial software supplier that aims to achieve a successful, profitable business by satisfying customers with high-quality software (or systems) products delivered on time and within budget. Although there are many reasons to use COQ in an organization, RES used it to measure and demonstrate the effects of their ongoing software improvement program.

Specifically, CoSQ was used to show the costs and benefits of their investments over the chronology of their improvement program. This is a good example of how a few leading organizations are using CoSQ.

RES and Its Improvement Program

RES develops real-time, mission-critical embedded software systems under contract for defense and commercial customers. They build these systems across air traffic control, vessel traffic management, digital communications, ground and shipboard radar, satellite communications, undersea warfare, military command and control, and combat training. The systems they build are large and typically range in size from 70 to 500 KDSI (kilo [1,000] delivered source instructions).

Since 1988, RES has been engaged in a software improvement initiative driven by the need to prevent schedule and budget overruns and by the crisis-driven environment that ensued. Once budgets and schedules were brought under some control, they turned their attention to reducing rework. The RES improvement initiative affected roughly 350–600 professionals.

The initiative's investment has been steady at about $1 million per year. An approach based on Carnegie Mellon University's Software Engineering Institute (SEI) Capability Maturity Model Integration (CMMI) approach was later adopted to focus their improvements. Following this approach, they were self-assessed at CMMI level 1 in 1988, at level 2 in 1990, at level 3 in 1992, and, as of 1995, operate all new projects starting at level 4. (For a description of the current SEI CMMI, see Chapter 7.)

Their improvement program strategy focused on both product and process. The product focal point areas included system definition, requirements definition, inspections and integration, and qualification testing. The focal point areas of the process included development planning and management controls, training, and pathfinding.

CoSQ

CoSQ was selected as one of four measures to track because it provided a framework for assessing rework levels and the overall ROI of the improvement program. The other measures chosen were software productivity, cost performance index, and overall product quality. The following describes how RES used CoSQ as a method for tracking rework and calculating ROI in their software improvement program. It is adapted from previously published sources.

The RES CoSQ Model

CoSQ was considered the sum of two components: the cost of nonconformance (waste or rework costs) and the cost of conformance (investment and monitoring costs). RES determined that they needed to track these quality cost categories. The model shown in Figure 8.2 was used to accumulate them.

In this model, CoSQ is the sum of costs associated with monitoring-, investment-, and rework-related waste. The focus on rework-related waste is essential because it constitutes a significant component of waste and scrap in software development. One of the costliest activities in software development today is rework. There are many underlying reasons for this, but most scrap and rework result from flawed requirements, changing conditions, and unexpected problems. For some reason, software professionals have come to accept mountains of rework as part of their everyday work—in some cases, they do not consider it rework. For example, valuable time is spent on fixing code defects when software doesn't perform as expected or on redesigning a user interface because the customer expects something different from what is provided. Many software engineers think of this as the natural way things are meant to happen.

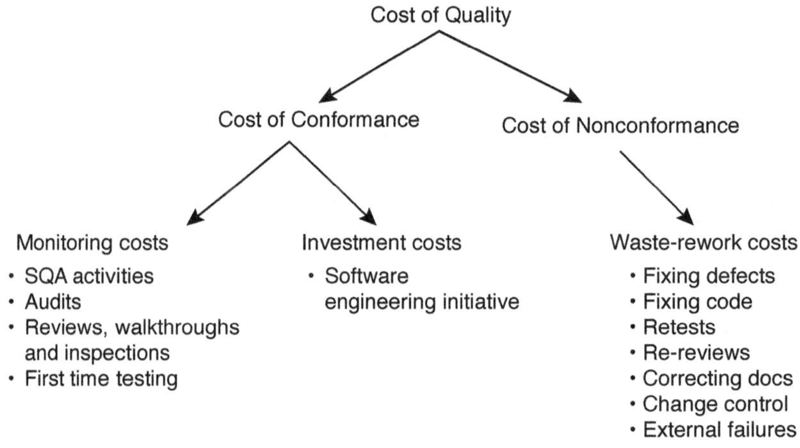

Figure 8.2 The RES CoSQ model.

The definitions of each subcategory, which were kept brief for simplicity, were subject to misinterpretation. Misinterpretation was addressed by refining the definitions as experience with them was gained. This required five iterations of the initial data-gathering exercise before a satisfactory level of consistency in definitions was achieved. Breaking these basic quality cost categories (waste due to rework, monitoring, and investment) into subcategories, defining the subcategories, and assigning project activities and costs to them proved to be a difficult task for RES because the existing work breakdown structure (WBS) used on the software projects did not correspond well to the CoSQ categories.

CoSQ Data Gathering

Project costs were collected using the conventional WBS, and project leads would periodically manually reclassify all costs to the COQ subcategories.

Project CoSQ data was then combined. Improvement initiative costs were factored in as a separate investment for the prevention project. These were then used to produce the organizational CoSQ average and trend data.

In the long term, RES wanted to develop a common work breakdown structure to provide the closest possible mapping to the COQ. This would also entail revisions to the cost accounting and time card reporting systems.

Experiences and Lessons Learned

CoSQ model usage lessons: RES encountered several experiences using the CoSQ model. Many questions arose about how to allocate costs to subcategories. There was variation in the methods used to allocate actual costs to the defined cost bins. This was resolved by refining the subcategory definitions and by analyzing and comparing the suballocation algorithms employed by the six target project leaders. It was necessary to have the project leader generate the data rather than an administrator because the project leader had firsthand knowledge of the project particulars and sound engineering judgment.

Using CoSQ Data to Understand the Impact of Improvement

Table 8.7 shows how RES used the CoSQ data to track the impact of their improvement program over the years. In 1988, RES began at CMMI level 1 and introduced its software process improvement (SPI) program. Using results from fifteen projects, they achieved CMMI level 3 practices in less than three years. As shown in Figure 8.3, at the level 1 stage, the RES CoSQ fluctuated between 55 and 67 percent of total project costs, and by the time the level 3 process maturity was reached, the CoSQ had dropped to approximately 40 percent of total project cost. In 1990, when RES was approaching CMMI level 3, total CoSQ accounted for approximately 45 percent of total project costs, and the ratio of conformance to nonconformance costs was approximately 3:2. In 1994, when RES adopted a goal of CMMI level 4, total CoSQ was approximately 24 percent of total project costs, and its ratio of conformance-to-nonconformance costs was approximately 3:1.

Table 8.7 CoSQ tracking at specific points in time.

Year	Performance	Waste-rework (Nonconformance)	Monitoring	Conformance	Investment
1988	34%	44%	15%		7%
1990	55%	18%	15%		12%
1992	66%	11%		23%	
1994	76%	6%		18%	

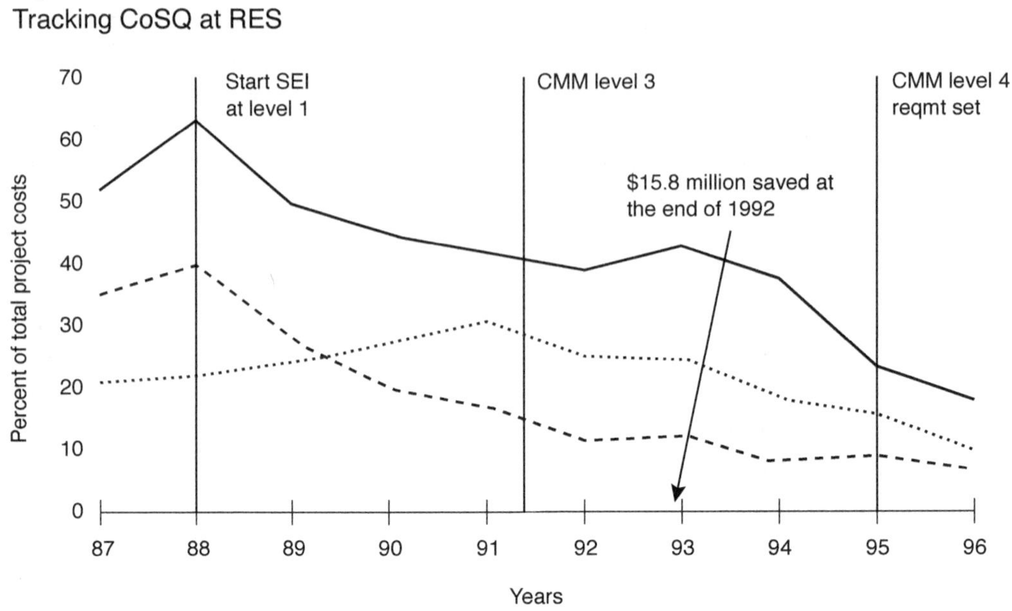

Figure 8.3 Tracking the cost of software quality at RES.

Rework Cost Savings

Figure 8.3 shows the trend in the average rework cost from the start of the improvement initiative. In the two years preceding the initiative, rework costs averaged approximately 41 percent of total project costs. In the two years that followed, the value declined to approximately 20 percent, and the trend continued downward. In 1995, CoSQ accounted for approximately 25 percent of total project costs, and rework due to both waste and downstream consequences was reduced to about 6 percent. Rework savings were achieved at the expense of a slight increase in monitoring or investment costs. For example, monitoring costs increased when formal inspections and investment costs replaced informal reviews and when inspection training was instituted. Additionally, rework costs associated with fixing defects identified during design increased from approximately 0.75 percent to approximately 2 percent of project cost, and those associated with fixing defects identified during coding increased from approximately 2.5 percent to approximately 4 percent of project cost.

The significant reduction in rework costs was attributable to fixing source code problems identified during integration, which fell to approximately 20 percent of the original level. The second-largest contributor to the rework reduction was the cost of retesting (also waste or rework), which decreased to approximately half its initial value. This clearly indicates that the additional costs of conducting formal inspections and the training required to precede them are justified by earlier identification of problems, resulting in more efficient integration.

Software Quality

The ultimate measure of quality is the contribution that software has made to RES's success with software-intensive systems. The improvements made have enabled success on several major software-intensive programs and have enabled RES to undertake larger software projects. This was demonstrated concretely on several complex system projects by removing software from the critical path and delivering early, thereby earning incentives. The primary quantitative measure that RES uses to assess overall product quality is the defect density in the final software products. This density factor is measured as the number of software trouble reports (STRs) per one thousand lines of delivered source instructions (KDSI) on each project. The project densities are combined to compute a monthly weighted average, yielding a time-variant plot of the trend. As shown in Figure 8.4, the average level of quality improved from about 17.2 STRs/KDSI to about 4 STRs/KDSI, about a fourfold improvement.

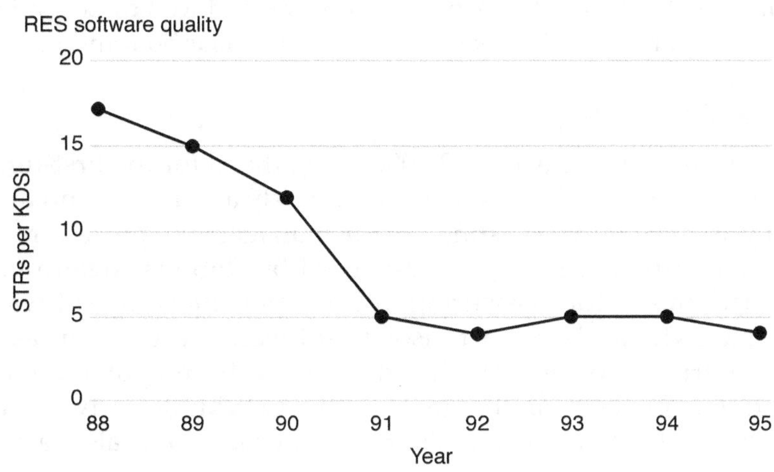

Figure 8.4 Tracking the level of software quality at RES.

Productivity

Data was collected from individual productivity projects on the equivalent delivered source instructions (EDSI) per person-month of development effort. The data was combined from all projects using a weighting function, and the results showed that average productivity was, in fact, increasing over time—meaning that jobs were costing less. Overall, RES achieved a 170 percent increase in software productivity across twenty-four projects over seven years.

Predictability

Management needed to be reassured that the improved productivity figures were being used to bid for new jobs. This issue was addressed by collecting cost performance index (CPI) data on the project's budgeted (predicted) cost and actual cost at completion (CAC). This cost performance index ratio (CAC/budget) for each project was then used to compute the monthly weighted average (using the same approach as time-variant COQ) to yield a plot of this time-variant measure. The results were encouraging, showing that the CPI had improved dramatically from approximately a 20 percent overrun range before the start of the initiative to a 1–2 percent range by early 1993. Overall, their CPI (CAC/budget) declined from approximately 1.43 to 1.00 over the first three years and has remained steady at approximately 1.00 since.

Other ROI Results

RES has won additional business based on process maturity results (no amount reported). Software personnel work less overtime than before, and this has led to lower turnover rates. The overall payoff of their improvement program is reported to be 7.5 times (not including a $9.6 million schedule incentive award in 1991). For example, in 1990, they spent $1.1 million on improvements and determined that the cost of nonconformance was reduced by $8.2 million in that year. Other leading organizations have realized similar bottom-line benefits.

CoSQ Costs and Benefits

CoSQ analysis cost RES about $25,000 of overhead for the first-time exercise. They repeated the CoSQ analysis approximately a year later and incorporated the analysis into the preexisting senior management process reviews on a semiannual basis. The CoSQ analysis used by RES was determined to be a viable mechanism for measuring the overall effect of software process improvement. It can be used to isolate software waste/scrap as part of an attempt to drive it to zero. The insights gained from applying the approach benefited the projects under analysis by providing early feedback. The improvement of the organization's standard process was also facilitated.

Institutionalization of CoSQ Tracking

By 1993, after three years, the data-gathering exercise for CoSQ had become routine. Although the complete analysis was conducted semiannually, some department managers required their project leaders to provide the CoSQ data along with their normal monthly tracking data. In retrospect, more emphasis could have been placed on transitioning to a common work breakdown structure geared to the collection of CoSQ.

Case Study Implications

Because RES used the CoSQ approach, the company learned many lessons that others will not have to learn the hard way. CoSQ was primarily used as an ex post measure of the SPI program's costs and benefits rather than as a tool to guide the SPI program (i.e., the CMMI). It is expected that other organizations using the CMMI approach to SPI will likewise do so. A detailed analysis of CMMI versus CoSQ will be necessary to focus investment opportunities on high-impact areas more effectively. This analysis should focus on the impact of prevention activities across the eighteen CMMI key process areas on internal and externally induced rework.

RES encountered specific difficulties in the following areas:

- when and how the CoSQ data was gathered, analyzed, reported, and used,
- how the model clashed with other models that were already in use,
- the lack of an existing WBS with well-defined categories that correspond easily to CoSQ,
- how the model was defined at the detailed levels,
- how the model was implemented in the organization,
- how CoSQ was used for root cause analysis, and
- how the model was used to stimulate SPI and quality improvements.

These difficulties can be overcome with appropriate training and coaching. RES's use of CoSQ occurred within the context of its contract-oriented systems business; therefore, the CoSQ approach was adapted to the specifics of that context. It will likely be necessary to modify the approach for use in other situations, with different business success parameters. These situations include

- standard product-oriented businesses,
- service delivery situations, and
- technology-based start-ups.

CASE 5: TRANSMOTOR, MANUFACTURING QUALITY COST ANALYSIS EXAMPLE

Reference

Juran, J. M., and Frank M. Gryna. *Quality Planning and Analysis: From Product Development Through Use*. 3rd ed. McGraw-Hill, 1993.

About the Operation

In one year, sales from the Transmotor Division of PDQ Company totaled approximately $25 million, with about 90 percent from industrial customers and 10 percent from government contracts. Profits after taxes were $1.2 million. Sales increased steadily from $1.5 million in January to $2.6 million in December. This increase was attributable to the introduction and widespread adoption of a new product design. The latest product was more reliable and cost less to produce. With a sizable amount of the new product in inventory at the start of the last year, the production rate was not increased until the second quarter.

During the year, a recently hired quality engineer began analyzing the quality program. He improved systems and procedures. Since the middle of last year, high rejection rates for the new product (both at final assembly and in parts) have required him to spend most of his time addressing the problems causing them: rejection rates.

The division's quality manager, Carl, has learned about the quality cost management technique and wants to determine whether it can benefit his division. Carl has attended several ASQ conferences and seminars and has spoken with quality control managers from similar companies. It appeared to him that quality costs, as a percentage of net sales billed, are typically 4–6 percent across companies producing similar products. He is not sure, however, which cost elements are included in his competitor's quality costs. A rough calculation of his division's costs for the prior year's month of October is shown in Table 8.8.

Table 8.8 Rough calculation of divisional quality costs.

Investment	$1,000
Monitoring	$100,000
Waste	$36,000
Downstream consequences	$27,000
Supplier quality cost	$164,000

Carl's first attempt at establishing a quality engineering program began over a year and a half ago with the hiring of an ASQ-certified quality engineer. Improvements in inspection methods and solutions to several chronic quality problems have since enabled Carl to reassign several inspectors and cover the increased production load without increasing the number of inspectors over the last three-quarters of the previous year. At this point, a considerable amount of 100 percent inspection remains; however, Carl believes that more inspection can be eliminated by improving manufacturing efficiency.

The cause of the current high rejection rate for the new product is not really known, and there is considerable finger-pointing. Manufacturing attributes the fault to a faulty design and the purchase of out-of-specification material. Design engineering reports that the existing tolerances are not being met and that parts are being mishandled before they reach the assembly area. Carl decides to identify which departments are the major contributors to high costs by implementing a quality cost program.

Starting the Program

The first decision the quality manager must make is which unit to include in the study. Since there is no breakdown by profit center within the division, the entire division is included.

Next, the quality manager discusses the concept and proposed program with the controller and requests aid in the initial study and future reporting. The controller is skeptical of the program but agrees to provide costs for those elements compiled and used for other purposes. The controller also agrees to provide personnel to assist in compiling other element costs as needed, provided it does not interfere with their regular duties.

The elements to be studied must now be selected. The elements shown in Table 8.9 are chosen as those most representative of the Transmotor Division operation. It is found that separate accounts are not maintained for specific elements, and estimates must be made for those items. In some cases, this requires splitting amounts in a general account according to an estimated fraction of the account attributable to the element. Some estimating can be done by counting the number of people performing such tasks as rework and sorting. Work sampling is also a valuable technique for such estimating. After determining the cost sources to use for each element, a detailed first study can be made. For the Transmotor Division, it was decided to collect data for the entire preceding year. This data is shown in Table 8.9. The actual costs for each category are plotted in Figure 8.5.

Waste, monitoring, and total costs exhibit an upward trend, as would be expected during a period of increasing activity. Investments in prevention haven't changed, but downstream consequence costs peaked in the first half of the year and appear to be leveling off. The next step is to identify appropriate measures of business activity (bases) to which the data can be related. The quality manager selected three possible bases:

- Sales base—net sales billed
- Cost base—cost of units shipped
- Labor base—factory hours

This data was collected from accounting and industrial engineering, and costs for each category were expressed as a percentage of the chosen base. Graphs of these percentages are shown as Figures 8.6, 8.7, and 8.8, respectively.

When expressed as a percentage of sales, total costs, monitoring, and downstream consequences decline, whereas investments in prevention and waste remain essentially unchanged. These trends, however, are not valid, as sales data is not a reliable measure of the activity that generates the costs. Most of the product billing originated from warehouse stock during the first half of the year.

Except for a December spike in quality costs as a percentage of factory hours, the ratios using factory hours exhibit trends similar to those using costs per unit shipped. (The December rise was due to a dip in factory hours during the holiday period.) Total costs peaked about midyear and appear to be decreasing. Monitoring and downstream consequences cost ratios exhibit a slight downward trend. Investment in prevention has not changed. This analysis highlights the rising waste cost ratio. A closer examination reveals that the major contributors to the increase are rework, scrap, and remedial engineering. These three elements were responsible for 91 percent of the waste costs. The most significant dollar contributor is scrap, accounting for 50 percent of waste costs. Graphs of waste costs as percentages of the expenses of unit shipping expenses are shown in Figure 8.9.

This raises the question: Where are these costs being generated? The quality manager requested a breakdown of the three most significant dollar contributors to waste costs: scrap, rework, and remedial engineering. It was found that three

Table 8.9 Total quality costs—Transmotor Division.

Elements	Jan	Feb	Mar	Apr	May
Investment					
Quality planning	500	550	400	300	350
Data analysis and preventive action	500	500	600	700	650
Planning by other functions	600	400	700	750	700
Dev. of measurement and control equipment	-	50	-	-	-
Training	-	-	-	-	-
Other investments	200	250	250	200	375
Total investment expense	*1,800*	*1,750*	*1,950*	*1,950*	*2,075*
Monitoring					
Inspection and test: purchased material	5,200	5,000	5,950	4,920	5,900
Laboratory acceptance testing	925	925	925	925	925
Maintenance and calibration of equipment	3,840	3,840	3,840	3,840	3,840
Depreciation of capital equipment	695	695	695	695	695
Inspection labor	52,300	53,250	52,275	52,325	51,250
Testing	29,120	30,950	30,050	28,245	29,350
Checking labor	2,710	2,805	2,740	3,117	3,240
Inspection and test material	475	80	316	940	510
Personnel qualification	-	-	-	30	-
Total monitoring costs	*95,265*	*97,545*	*96,791*	*95,037*	*95,710*
Waste expense					
Scrap- division caused	25,170	15,025	19,112	18,997	28,040
Rework- division caused	5,200	6,150	6,210	4,925	9,010
The supplier caused losses	1,200	1,099	1,248	1,170	1,370
Troubleshooting	2,080	1,975	2,125	2,020	2,115
Remedial engineering	4,200	4,250	7,125	8,010	7,850
Total waste costs	*37,850*	*28,499*	*35,820*	*35,122*	*48,385*
Downstream consequences					
Product warranty	19,670	22,300	22,960	24,850	22,100
Returned product cost	1,800	1,800	1,800	1,800	1,800
Field service costs	7,100	7,100	7,100	7,100	7,100
Total downstream consequence costs	*28,570*	*31,200*	*31,860*	*33,750*	*31,000*
Total quality costs	**163,485**	**158,994**	**166,421**	**165,859**	**177,170**
Measurement bases					
1. Net sales billed	1,525,000	1,420,500	1,872,500	1,810,200	1,798,400
2. Factory hours	82,650	83,152	82,164	81,245	82,360
3. Cost of units shipped	1,225,000	1,315,500	1,275,250	1,095,650	1,080,975

Jun	Jul	Aug	Sep	Oct	Nov	Dec	Total
250	-	200	250	-	100	100	3,000
750	250	800	750	1,000	900	900	8,300
650	650	650	600	700	700	650	7,750
-	-	-	-	-	-	-	50
-	-	250	-	-	-	-	250
190	750	260	460	225	190	220	3,570
1,840	*1,650*	*2,160*	*2,060*	*1,925*	*1,890*	*1,870*	*22,920*
6,010	3,900	6,410	7,125	6,500	6,400	7,450	70,765
925	925	925	925	925	925	925	11,100
3,840	3,840	3,840	3,840	3,840	3,840	3,840	46,080
695	695	695	695	695	695	695	8,340
53,200	48,875	51,450	52,050	52,725	51,400	50,575	621,675
31,940	30,125	35,830	35,750	38,700	43,525	44,100	407,685
3,120	3,250	3,325	3,390	3,470	3,515	3,570	38,252
425	270	317	430	525	130	100	4,518
-	-	-	-	30	-	-	60
100,155	*91,880*	*102,792*	*104,205*	*107,410*	*110,430*	*111,255*	*1,208,475*
33,980	9,060	20,050	22,150	18,220	27,110	24,140	261,054
6,020	7,800	10,500	12,250	10,875	12,900	12,040	103,880
2,715	1,110	1,795	1,745	1,890	1,375	2,160	18,877
2,170	2,050	2,265	2,450	2,645	2,725	2,945	27,565
9,100	10,460	13,610	12,990	13,060	11,550	13,510	115,715
53,985	*30,480*	*48,220*	*51,585*	*46,690*	*55,660*	*54,795*	*527,091*
20,990	20,500	19,550	18,850	20,110	18,900	19,750	250,530
1,800	1,800	1,800	1,800	1,800	1,800	1,800	21,600
7,100	7,100	7,100	7,100	7,100	7,100	7,100	85,200
29,890	29,400	28,450	27,750	29,010	27,800	28,650	357,330
185,870	**153,410**	**181,622**	**185,600**	**185,035**	**195,780**	**196,570**	**2,115,816**
1,896,750	2,086,550	2,314,640	2,402,500	2,276,550	2,697,540	2,625,400	23,201,530
91,200	83,750	96,750	112,750	115,750	115,700	91,250	1,118,721
1,205,620	1,125,050	1,397,450	1,334,150	1,400,500	1,602,930	1,625,625	15,683,700

sections of the shop (winding, feeder 1, and feeder 5) generated 82 percent of the scrap during the previous year.

Two sections (winding and assembly) contributed 73 percent of the rework charges, and one model (Model T) accounted for 60 percent of the remedial engineering. Figure 8.10 shows Pareto graphs of these cost breakdowns.

The quality manager now has sufficient information to begin developing a corrective action program. To be effective, it must involve both the manufacturing manager and the engineering manager, since the elements that are the most significant contributors to waste costs are under their respective functional managers' responsibility. The quality manager must also examine costs that are his own responsibility—investment and monitoring. Reductions in overall expenses will require a program of cooperative effort.

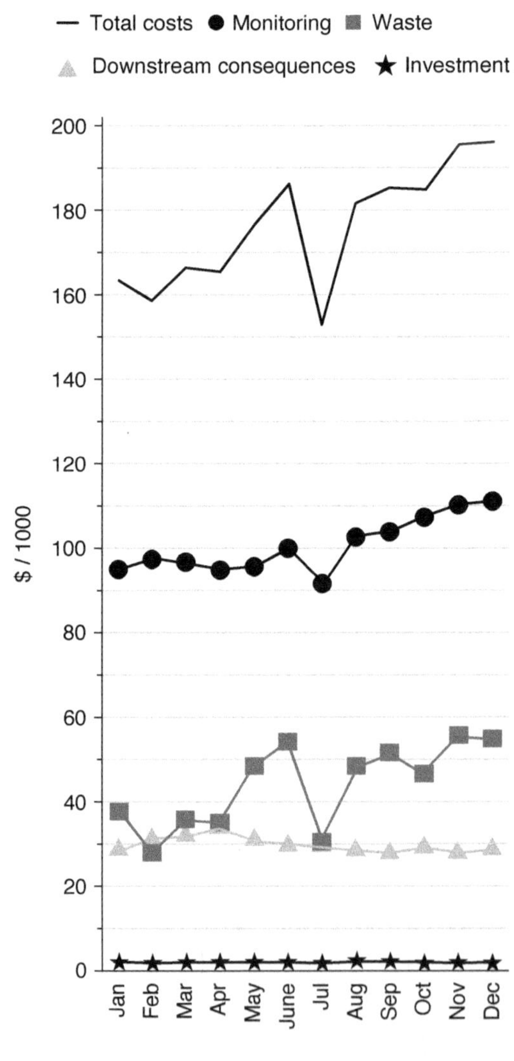

Figure 8.5 Actual quality costs.

A meeting is held with the responsible functional managers, at which the quality manager presents the results of his study and analysis to date. All managers are asked for their views on the program, whether the costs for which they are responsible could be reduced, and if so, by how much. At this meeting, objectives and plans for their achievement are solicited from each responsible manager.

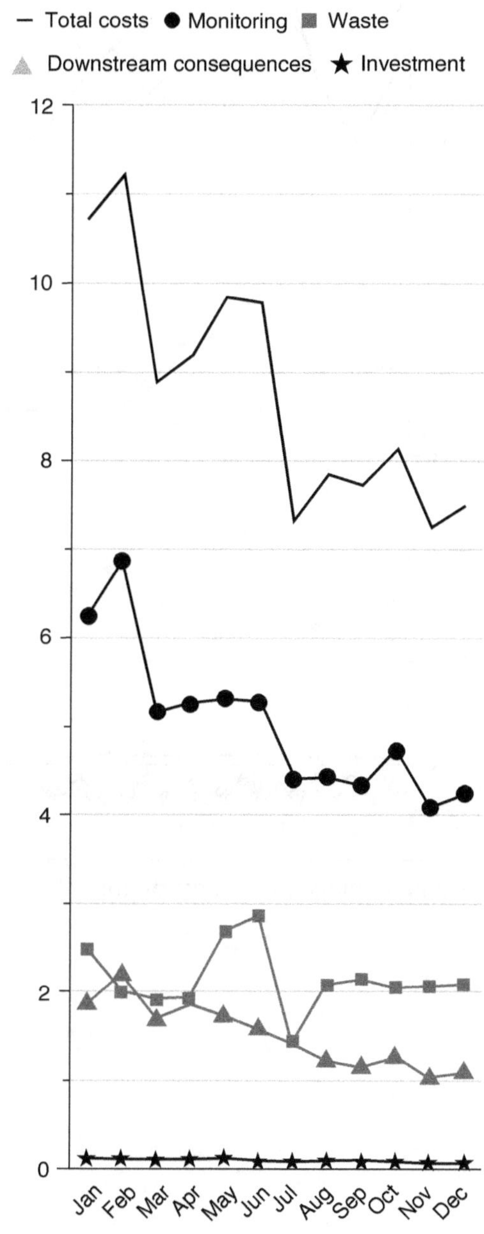

Figure 8.6 Quality costs as a percentage of net sales billed.

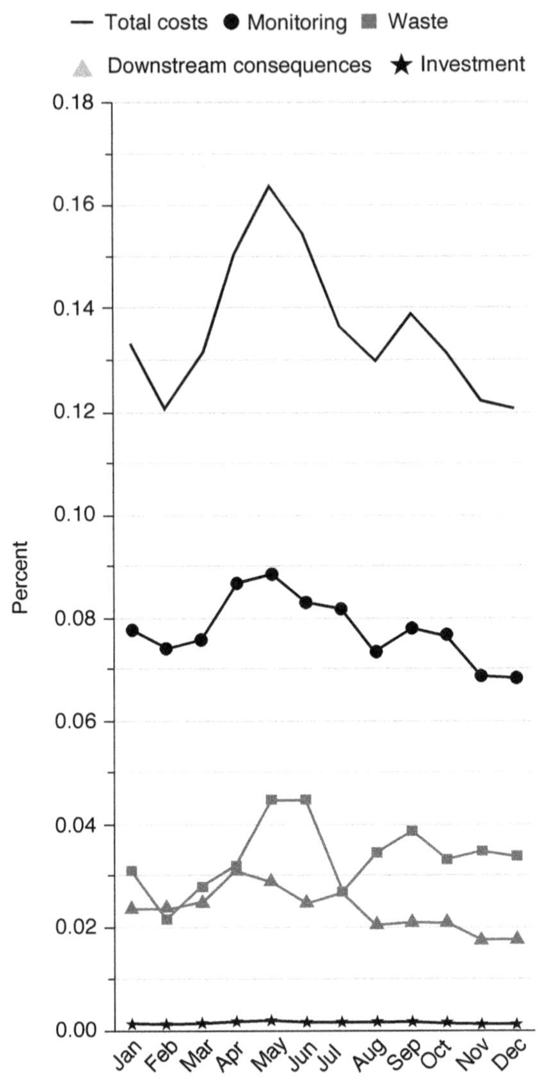

Figure 8.7 Quality costs as a percentage of the cost of units shipped.

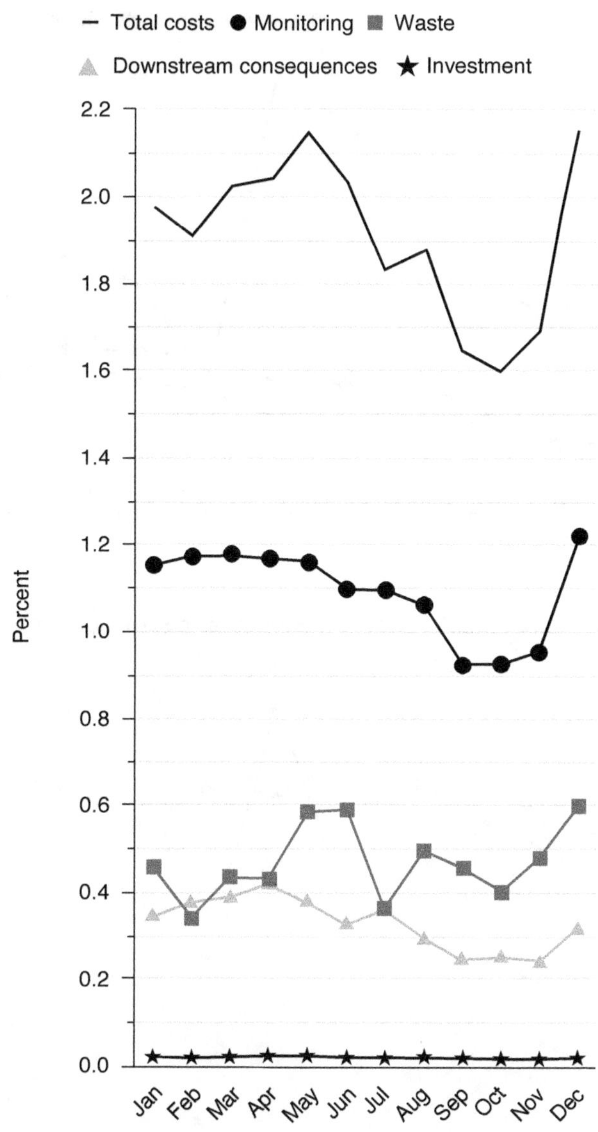

Figure 8.8 Quality costs as a percentage of factory hours.

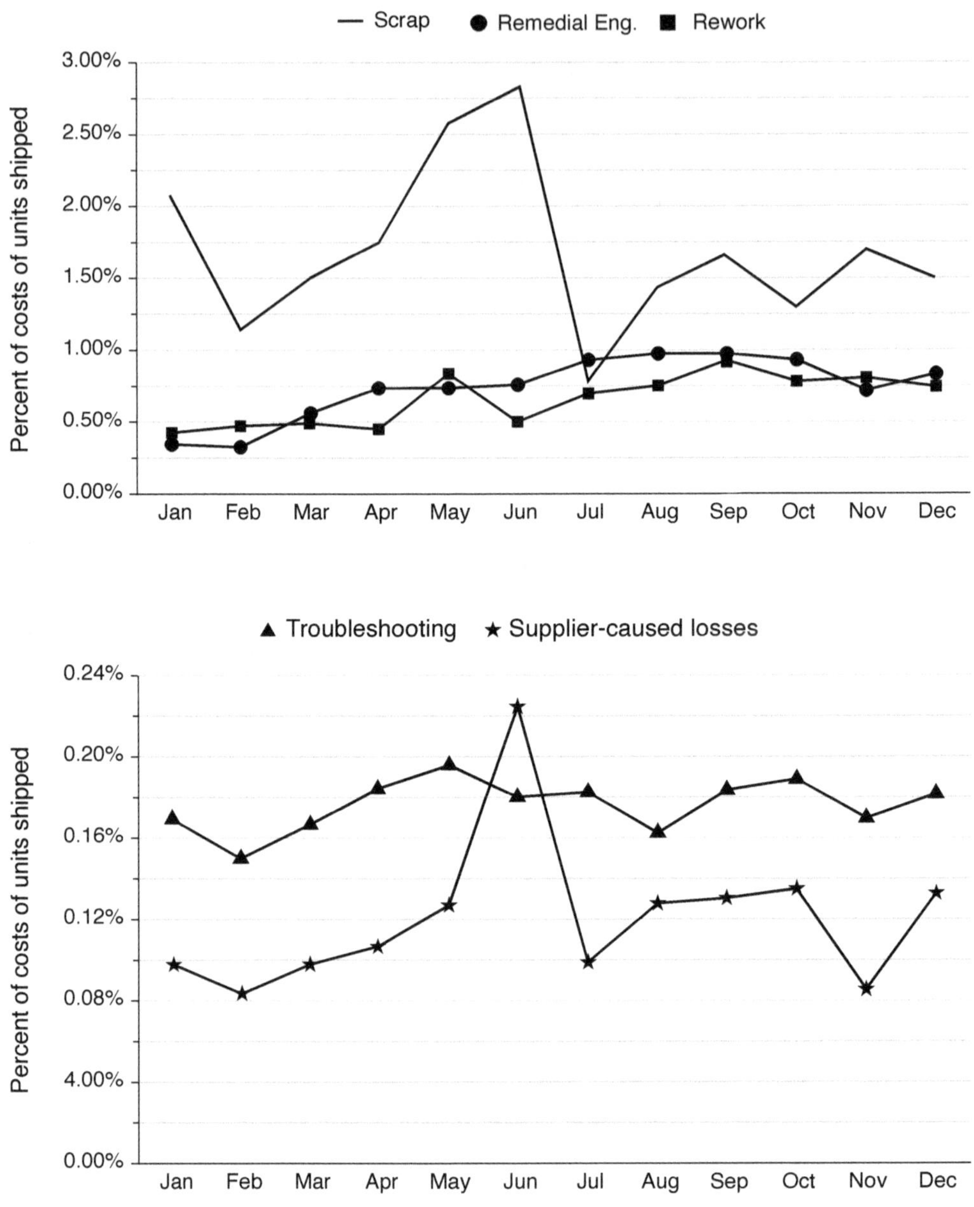

Figure 8.9 Waste costs.

The initial report contains all data and graphs generated in the study, the objectives for cost reduction, and the plans for achieving them. Objectives developed by the management team are as follows:

- Reduce the overall quality costs to 6.7 percent of net sales billed within one year.
- Reduce monitoring costs to 3.3 percent of net sales billed within two years and to 3.6 percent within one year.
- Reduce waste costs to 1.8 percent of sales within one year.

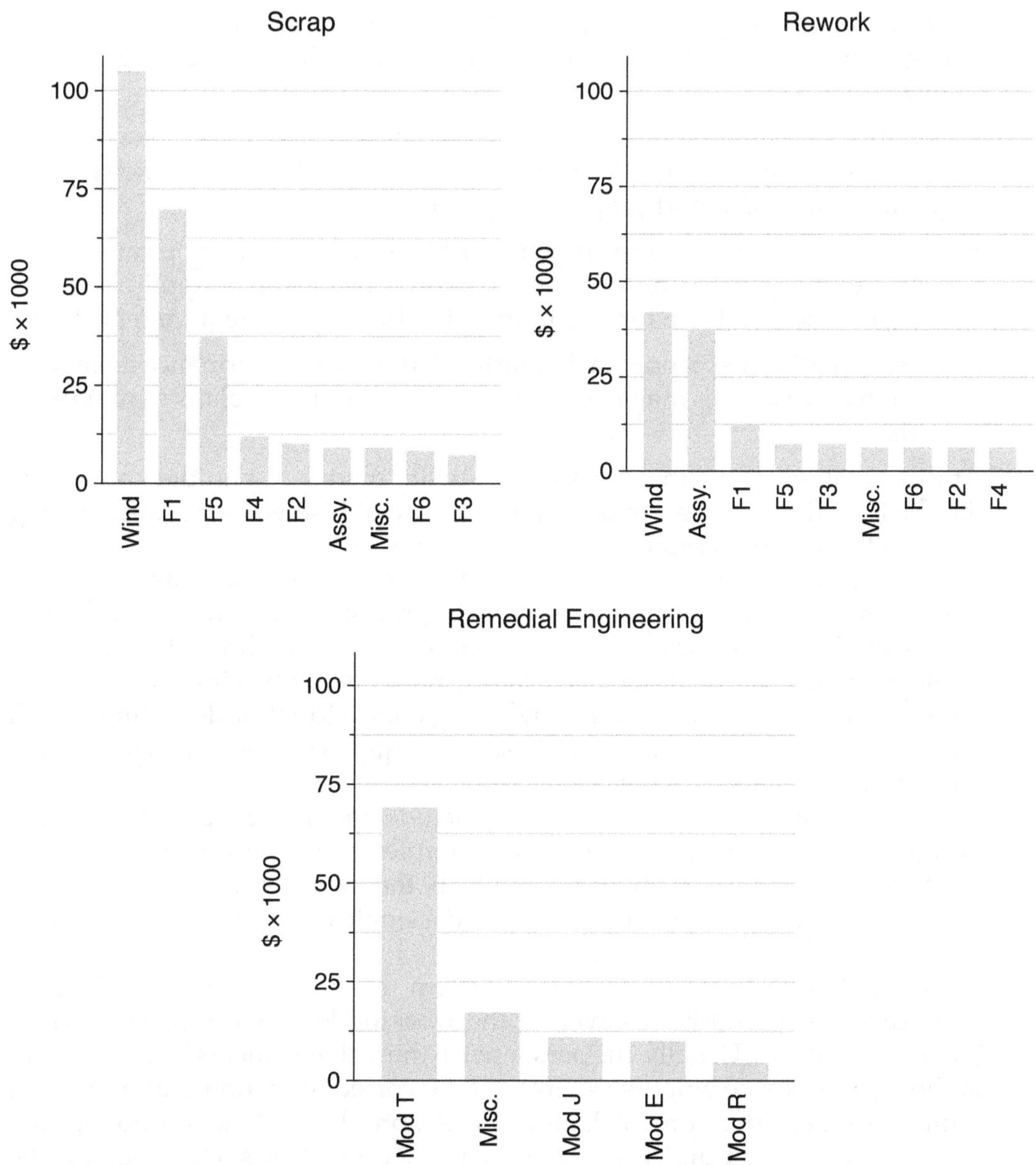

Figure 8.10 Pareto distributions of scrap, rework, and remedial engineering.

- Maintain downstream consequence costs at no more than 1.1 percent.

Action programs for the attainment of these objectives are as follows:

- Hire an additional quality engineer charged with the responsibility for the identification of the causes of high-cost problems and the coordination of corrective action.
- Reassign the present quality engineer to the task of finding less costly ways of inspection and test and ways to eliminate 100 percent inspection.

- Set up teams comprising representatives from manufacturing, engineering, and quality in sections of the shop with high scrap and rework costs.
- Set up a Six Sigma task force to determine the cause of the excessive remedial engineering costs on the Model T unit. Establish an objective to reduce costs and report progress regularly.
- Determine the causes of the highest cost field problem. If shop caused, assign to the shop teams for action. If caused by design or application, assign to the task force for attention. Set objectives and require reporting.
- Issue a quality cost report each quarter that shows performance against objectives, discusses major problem areas, and outlines plans to reduce costs.

At this point, we leave the Transmotor quality manager with the knowledge that he is well on his way to eliminating many of his headaches and making the quality function a profit contributor for the division.

Once a problem has been identified and reported and the relevant personnel are committed to action, the process begins, but it is far from complete. The efforts of the people involved must be planned, coordinated, scheduled, implemented, and monitored. Typically, problems can be classified into two types: those that can be corrected by a single individual or department with little or no external assistance and those that require coordinated action across multiple activities within the organization.

Examples of the first type of problem include operator-controllable defects, design errors, and inspection errors. Examples of the second type include product performance problems for which the cause is unknown, defects caused by a combination of factors beyond a single department's control, and field failures of unknown cause.

The first type needs no elaborate system to attack and solve problems. Most can and should be resolved at the working level with the supervisor, engineer, or others. Usually, the personnel in these departments have sufficient authority (employee empowerment) to take corrective action within defined limits without prior approval. Unfortunately, problems of the second type are usually the costliest and least easily solved. Causes of such issues can be numerous and unknown. Solutions may require action from several sources.

The investigation of the problem and the planning of its solution must be coordinated and scheduled to ensure effective action. One of the best devices for doing this is the Lean-Six Sigma or quality improvement team. Working with the data and problem analysis reports, and headed by an individual interested in solving problems, this team develops the plan and then coordinates and schedules the investigation and action. It has been found that a curious individual with line responsibility makes the best head for such a group. Each project should be tracked. Meetings should be held regularly, and minutes should be published. The DMAIC methodology is a good one to follow.

CASE 6: PROFIT IMPROVEMENT FOR AN ELECTRICAL FIRM

Reference

Harrington, H. James. *The Poor Quality Cost*. Marcel Dekker, 1987.

A large electrical firm (the IBM San Jose facility) initiated a corporate program to identify, analyze, and reduce quality costs. It was called the Product Integrity Improvement (PII) program. A formal management commitment to improving quality and an organized approach have led to profit improvements across several divisions. This example describes the approach used and the results obtained in one location.

In most industrial environments, the greater amounts of total quality costs are found in waste and downstream consequence costs. Most organized efforts to reduce costs and improve profits are concentrated in this area. It should be stated, however, that by placing greater emphasis on investment in prevention activities, a significant improvement in quality costs will be realized. The PII program was intended to focus attention on all phases of quality costs and thereby improve the quality, safety, reliability, and environmental effects of products while reducing total quality costs.

Establishing and Implementing the Division PII Program

It was decided to implement the PII program in the example division to place greater emphasis on the total quality costs. The primary events that occurred to establish and implement this program were as follows:

1. Received top management commitment, support, and involvement.
2. Organized the PII program in the division.
   a. Assigned responsibility for the PII program to a member of the division manager's immediate staff.
   b. Established a PII council to assist the PII program manager in determining the overall approach, developing division strategy, and implementing the program. The council members included the division engineering manager, the manufacturing manager, the controller, and the quality manager.
   c. Conducted a PII seminar with headquarters quality assurance assistance to introduce the concept to responsible management personnel.
3. Identified the quality cost elements and selected account sources.
4. Collected all the quality costs for the division for the previous twelve months to establish the total quality cost baseline.
5. Analyzed the division quality cost data and identified the most significant quality cost expenditures. Analysis of the data indicated that waste costs

were requiring a disproportionate expenditure and should receive the highest priority for action.

As a result of this analysis, quality costs were found as shown in Table 8.10.

Table 8.10 PII quality costs.

	Percent of sales
Investment	0.1
Monitoring	1.5
Waste	2.3
Downstream consequences	1.5
Total	5.4

Waste costs were analyzed to find the high-cost contributors (see Table 8.11).

Table 8.11 PII waste costs.

Item	Approximate percent of total waste cost
Cores	5
Wire	20
Coil winding—assembly	35
Final assembly/test	15
Other	25

The "coil winding—assembly" area was selected for focused investigation since collectively it accounted for the most considerable portion of the total waste cost. The investigation and improvement program encompassed these steps:

1. Determined fundamental problems and underlying causes of the issues and assigned responsibility for corrective action.
   a. Identified three underlying causes:
      i. Operator winding errors
      ii. Damage to coils in handling
      iii. Design problems
   b. Established a quality improvement team:
      i. Manufacturing manager—chairperson

   ii. Quality manager
   iii. Manufacturing engineering (equipment problems)
   iv. Engineering (design problems)
   v. General supervisor (operator problems)
2. Established quality cost improvement objectives as an integral part of the division's profit plan.
3. Created reporting systems to provide accurate cost visibility and to measure improvement performance.
4. Met weekly to review progress, establish plans, and assign new tasks to be completed.
5. Reviewed monthly total quality costs against the objectives and initiated corrective action where needed.
6. Educated, trained, and emphasized the importance of everyone doing the job right the first time. Employee involvement was most important in attempts to achieve improvement. This was accomplished in several ways:
   a. Established meetings with the general supervisor, quality supervisor, and section supervisors to
      i. Identify key projects to be worked on
      ii. Plan programs for improvement
      iii. Review progress
   b. Developed defect charts for each manufacturing section, showing objectives and actual costs
   c. Conducted workplace meetings to establish a quality-conscious attitude
   d. Set up training programs for certain critical skill, high-cost areas
7. Recognized individuals and/or groups that made significant contributions toward improvement.

Results obtained from this program were significant (Table 8.12).

Table 8.12 Results from PII: quality costs as a percent of sales.

	Previous year actual	Current year objectives	Current year actual
Investment	0.1	0.2	0.2
Monitoring	1.5	1.3	1.2
Waste	2.3	2.0	2.1
Downstream consequences	1.5	1.1	1.2
Total	**5.4**	**4.6**	**4.7**

Summary of the PII Program

The success of a PII program depends upon many factors:

- top management involvement and support,
- visible total quality cost data,
- the setting of division objectives for improvement and the monthly reporting of performance against objectives,
- organization for improvement,
- the establishment of a quality improvement team, and
- employee involvement and recognition.

The benefits to be gained from a properly implemented PII program include

- reduced total quality costs with a corresponding increase in profits,
- improved product performance, product integrity, and adherence to schedule,
- increased customer acceptance of products and services, and
- increased repeat sales and new sales from improved product reputation.

CASE 7: DEPARTMENT OF DEFENSE (DOD) CONNECTOR WASTE FAILURE COST IMPROVEMENT

Reference

Harrington, H. James. *The Poor Quality Cost*. Marcel Dekker, 1987.

A major connector manufacturer (ITT Cannon) elected to enter the connector market with a new connector design to meet DoD requirements. The following discussion illustrates how, over four years, the product assembly waste cost was reduced from $180,000 per year to $20,000 per year.

History

During the program's early phases, typical start-up problems associated with new manufacturing techniques, planning, and training occurred. These items were resolved one by one until assembly waste costs decreased to approximately $180,000, a figure still considered above normal for a new connector program.

Three years later, a waste cost plateau of $150,000 per year had been reached, a figure that was still comparatively high. It was not known that a considerable portion of waste dollars originated in the assembly departments. Using the available financial data, waste costs by specific cost center in the assembly departments were identified. (Each cost center is responsible for the assembly of a particular type of connector.)

It was observed that two specific cost centers accounted for more than 75 percent of the assembly departments' total waste costs during that period. One major cost center was the manufacturing of bonded assemblies for other connectors. This area was also the target for significant cost improvements.

Using the first half of the year as a base and holding sales constant in the second half, it became apparent that another large-dollar waste year would result in the connector area (Table 8.13). It was necessary to attack the problem immediately.

Approach

A review of the prior six months' waste history was initiated. All discrepancy reports were evaluated, summarized, and categorized, and high-scrap and rework-cost areas were identified (Figure 8.11). It was apparent from this summary that nearly 85 percent of all rejections resulted from parts contaminated by improperly or excessively applied adhesives. Because of the labor required to remove the adhesives, on average, eighty connectors per month were scrapped. This alone amounted to almost $20,000 per year.

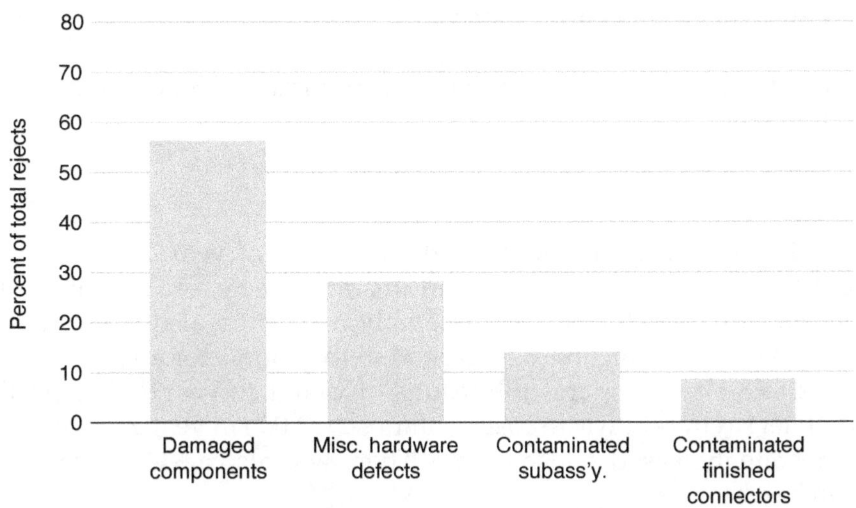

Figure 8.11 Rejection causes.

Causes

Based on this data, these causes were identified:

1. The methods used to apply adhesives for this new connector series were outdated.
2. Operators required additional training in applying adhesives.
3. Operators and inspectors did not understand the complete workmanship standards developed for this program. As a result, the operators were performing unnecessary rework.

As a result of these causes, some operators were producing a higher percentage of rework and scrap than others.

Corrective Measures

Once the causes were identified, a corrective action team was formed to include a quality engineer, an industrial engineer, and the production supervisor. The production supervisor was appointed to head the team. They accomplished a number of significant changes:

1. They reviewed manufacturing instructions, revised methods, obtained new tools, and revised workmanship standards.
2. They introduced a new method for applying adhesives using a silk-screen technique.
3. They retrained or replaced the operators causing the most rework and scrap.
4. They retrained inspectors and operators in the requirements of the modified workmanship standards.

The costs for these changes were approximately $4,000 for tooling and $3,000 for labor.

Results

Table 8.13 compares the results of the first half of year 1 with those of the second half of year 1. Waste costs in this area decreased dramatically. Note that waste dollars in the bonded assembly cost center also reduced as a result of the application of techniques learned in the connector area.

Table 8.13 also compares the results of year 1 to year 2. The results have yielded net savings in waste costs of almost $80,000 in year 2 for the connector area, despite increased sales volume. There was also a $34,000 savings in the bonding area.

Table 8.13 Waste by cost center.

	CC:241 (Bonding) ($ 000)	CC2450 (Connector assembly) ($ 000)
Year 1 first half	57.6	67.0
Year 1 second half	17.7	34.0
Year 1 total	75.3	101.0
Year 2 total	41.2	22.0
Net Savings	34.1	79.0

Summary

Using the available tools, the team identified a problem, determined its cause, and initiated corrective action to prevent recurrence. This resulted in a net savings equivalent to a sales increase of over $350,000 on these products. This was accomplished by recognizing that small pieces of a larger problem can more easily be digested and resolved one at a time.

CASE 8: BANKING

Reference

Sambamurthy, V., and Robert W. Zmud. "Banc One Corporation: Measuring the COQ." *IT Management Case Study Series*. 1994 by Financial Managers Society (FMS).

Eleven business units of Banc One Corporation participated in a study of quality costs, customer satisfaction, and quality deficiencies/defects. The relationship between these three measurable components of quality was explored. Correlation analysis was performed using these three metrics. Interesting relationships emerged that can help predict what you will find and improve in an organization if each metric is measured over time.

Attacking high COQ opportunities reduces costs, defects, and errors and significantly increases customer satisfaction and delight. A COQ analysis helps prioritize the opportunities for improvement. This prioritization process ensures the selection of opportunities with the greatest impact on the bottom line. The quality improvement team's processes for addressing COQ opportunities are discussed (Figure 8.12). Banc One uses this process on an ongoing basis to allocate resources and to motivate or assign quality improvement teams to improve opportunities. In subsequent years, they have enhanced net income by $20 million annually through the application of these processes.

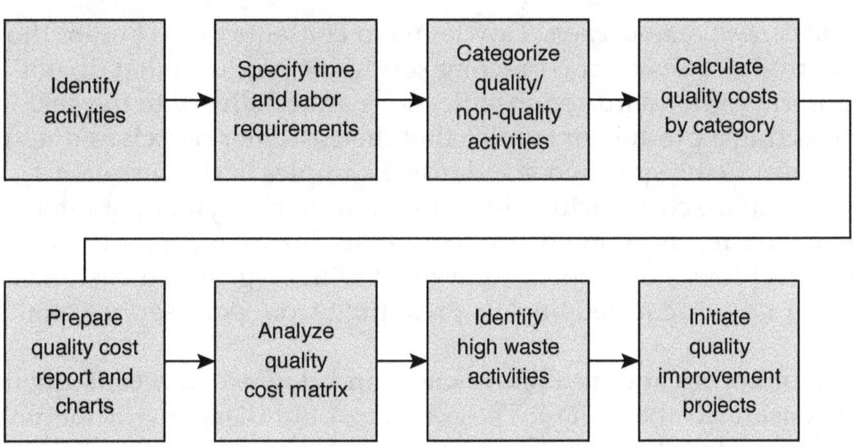

Figure 8.12 The Banc One quality improvement process.

Introduction

Within the service industry, as in manufacturing, there are costs associated with providing and ensuring high-quality services that are designated as quality costs. To understand the economic impact of quality costs on a bank's financial position and how they can be used to enhance that position, quality costs must be defined, measured, analyzed, and managed. The four quality cost categories of process investment, monitoring, failure waste, and downstream consequences, as they relate to the banking industry, are defined as follows:

- *Process investment.* Process investment costs are proactive activities undertaken before or during processing or service delivery. Process investment costs are those costs associated with operations or activities that prevent failures from happening and keep monitoring costs to a minimum. Examples of process investment activities are new product review, quality planning, quality improvement team meetings, training programs, written policies and procedures, analysis of quality information, and quality information/ improvement projects.
- *Monitoring.* Monitoring costs are those incurred to project or predict quality levels and to assess the condition of a product or service to determine its degree of conformance to quality standards or specifications. Monitoring activities encompass inspecting incoming work, supplies, and materials; conducting periodic in-process inspections; performing checks, balancing, and verification; carrying out final inspections; analyzing customer feedback via surveys and focus groups; and reviewing customer correspondence and complaints.
- *Failure waste.* Failure waste costs are those incurred to correct services or products that do not conform to standards or specifications before delivery to the customer. Examples of failure waste are machine downtime, scrap, waste due to improperly processed forms or reports, and rework of incorrectly processed work.
- *Downstream consequences.* Downstream consequence costs are those incurred as a result of correcting service or products that do not conform to standards or specifications after delivery to the customer or correcting a product or service that the customer perceives does not conform to the specified standards. Examples of downstream consequence costs include investigation time, payment of interest penalties, reprocessing of an item, scrap due to improperly processed or incorrect forms or reports, time spent with disgruntled customers, and lost or never-acquired business resulting from poor service quality.

Most monitoring and all failure waste and downstream consequence costs were considered the COPQ. These costs would disappear if the product or service were defect-free and conformed to standards and specifications at the time of completion or delivery.

When the four categories of quality costs are combined, the total quality cost for a product, function, or service is obtained. In aggregate, these costs range from 10 to 30 percent of sales and from 25 to 40 percent of operating expenses. The latter is how quality costs are measured in banks that do not have a sales function. This range has been verified in banking time and time again. It represents a tremendous opportunity to reduce costs and increase customer satisfaction.

Objective

The objective of a quality cost measurement is to aid a company in determining, by function, by product, or by product line, where the highest costs of quality are so that improvement efforts can be targeted there. The goal should be to improve quality, productivity, and profitability by reducing quality costs. Revenue can also be increased by improving customer satisfaction, which in turn drives loyalty and repeat purchases.

The quantitative data from quality cost measurement in supply management provides the necessary information to optimize quality improvement resource allocation. High monitoring and waste costs alert management to problems that may have been previously overlooked. As a result, improvement projects can be undertaken to reduce cost and improve quality, productivity, and profitability. A company-wide program alerts employees to the importance of being accountable and their responsibility to meet the company's quality objectives. The company-wide commitment to quality increases productivity, since work done the first time does not require redoing. Furthermore, increased productivity, higher profitability, and greater customer satisfaction should give the company a stronger position in the competitive marketplace.

Getting Started

The company's specific objective will determine how the process is to be implemented. However, there are basic guidelines to follow when implementing quality cost measurement:

- Start small.
- Do not attempt to quantify the cost of every product, service, or function within the company.
- Start with activities that you suspect may have high failure or monitoring costs.

Once you have selected the area to be measured, begin data collection. First, develop a complete list of all operations, activities, and jobs to estimate costs based on actual operating activity and expenses. The list details each task or operation performed in the area and the time spent performing it. Pro-investment activities and time spent performing them are included on this list since they are part of the total quality costs. Then, develop a flowchart to ensure that all activities are captured and to balance labor activities and expenses. Time can be allocated as a percentage of total available labor time or as a fixed number of hours (recommended). The total hours spent performing all tasks should equal the total labor hours available in the area. In either case, the total number of available labor hours in an area must be determined and specified, along with the average employee salary.

Categorize each as a quality activity (in one of the four categories) or a nonquality (production) activity. Once quality activities, salary level, and time spent on each activity have been specified, quality costs can be calculated by multiplying each activity's labor time by the average salary level.

Analyze the expense statement to determine whether any additional direct expenses are quality costs. If so, assign them to a specific quality cost category.

Examples of other direct expenses include

- internal or external training expenses charged (process investment),
- the purchase or maintenance of test equipment (monitoring),
- destruction of forms (failure waste), and
- interest penalties or income lost due to poor quality (downstream consequences).

List the loan activities and their related costs, by category, on the quality cost report and total it (Table 8.14). Only those activities that fall into the four categories are listed. (In Table 8.14, UCC refers to a Uniform Commercial Code filing, creating a legal lien on the personal property associated with that real property.) Then identify the most significant single opportunities for monitoring and waste. High monitoring costs indicate potential areas for improvement that can be reduced through effective prevention. If it's not practical to start all efforts immediately, develop a Pareto diagram using this data to determine which project to undertake first (Figure 8.13).

The success of a quality improvement project depends on accurate diagnosis and the application of cause-and-effect analysis to the problem at hand. The quality cost data specifies the symptom or effect; however, until the causes are known, identified, and tested, no action should be taken to correct the problem. Attempting to solve a problem without knowing the cause could increase quality costs and negate the purpose of quality cost measurement. Several good root cause analysis techniques can be used here. Having correctly identified the cause or causes of the problem, the team can use the improvement project to set goals and a course of action to correct it. Potential savings based on the proposed course of action can then be estimated. The team can evaluate the success of the improvement project and the quality cost program by measuring the actual shift or reduction in quality costs, along with internal quality levels and external customer perceptions. Typical improvement activities yielded returns of 5:1 or even 10:1 on investment. These eleven quality improvement projects yielded an average return of 8:1.

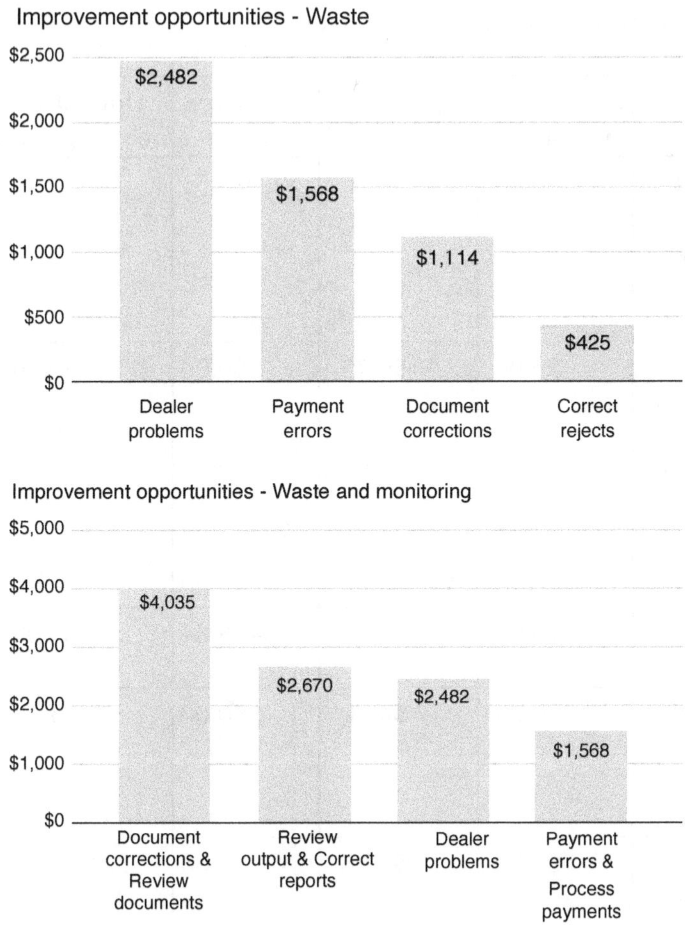

Figure 8.13 Pareto diagrams—waste and monitoring improvement opportunities.

Table 8.14 Quality cost report—installment loans.

Operation	Process investment	Monitoring	Failure waste	Downstream consequences	Total
Making a loan					
Run a credit check	$0.00	$0.00	$26.13	$0.00	$26.13
Prepare and put through General Ledger (GL) tickets and Income/Loss (I/L) input sheets	$0.00	$0.00	$248.19	$0.00	$248.19
Review documents	$0.00	$3,013.78	$7.84	$0.00	$3,021.62

(continued)

Table 8.14 Quality cost report—installment loans. *(continued)*

Operation	Process investment	Monitoring	Failure waste	Downstream consequences	Total
Make document corrections and gather additional documents or information	$0.00	$0.00	$1,013.65	$0.00	$1,013.65
Prepare tickler file; review and follow up on titles; insurance; second meetings and UCCs	$0.00	$156.75	$0.00	$0.00	$156.75
Review all output	$0.00	$2,244.14	$0.00	$0.00	$2,244.14
Correct rejects and incorrect output	$0.00	$0.00	$425.84	$0.00	$425.84
Work associated with the incomplete collateral report	$0.00	$0.00	$0.00	$73.38	$73.38
Work associated with dealer calls, dealing with any problems, and the time to research and communicate	$0.00	$0.00	$0.00	$2,481.88	$2,481.88
I/L system downtime	$0.00	$0.00	$519.89	$0.00	$519.89
Time spent training or being trained on I/L	$1,366.34				$1,366.34
Loan payment					
Receive and process payments from all sources	$0.00	$261.25	$783.75	$0.00	$1,045.00
Respond to inquiries when no coupon is presented with payments	$0.00	$0.00	$783.75	$0.00	$783.75

Operation	Process investment	Monitoring	Failure waste	Downstream consequences	Total
Loan payoff					$0.00
Receive and process the payoff and release document	$0.00	$0.00	$13.06	$0.00	$13.06
Research payoff problems	$0.00	$0.00	$13.06	$0.00	$13.06
Total COQ	$1366.34	$5,675.92	$3,835.16	$2,555.26	$13,432.68
COQ as a percentage of total quality cost	10.17%	42.25%	28.55%	19.02%	100.00%
COQ as a percentage of reported salary expense	2.60%	10.79%	7.29%	4.87%	25.55%

In addition to the team monitoring quality cost changes that were effective at the conclusion of the project, continuous measurement should be conducted to ensure the gain has been sustained.

Relationship of the COPQ to Defects and Customer Satisfaction

The results of eleven quality cost measurement studies included quality costs as a percentage of total expense, the overall defect rate from quality standards, the most significant individual quality improvement opportunity identified, and the overall customer satisfaction measure for each department or service. When these measurements were correlated, several interesting relationships emerged. Low, poor-quality costs as a percentage of total quality costs indicate a small opportunity for improvement. Conversely, a high percentage usually predicts a significant opportunity for improvement.

When defect rates are not correlated to improvement opportunities, neither high nor low defect rates predict improvement opportunities of the same relative size. In fact, a low defect rate can often obscure substantial opportunities for individual improvement.

The data in this study showed that where ample opportunities to improve individual quality costs exist, customer perceptions of the service level provided are unacceptable. Therefore, defect rates are not very useful in identifying or predicting the level of quality cost improvement opportunities.

Customer survey results may be a better indicator as to which areas will benefit from quality cost measurement and improvement team efforts. Once the problem has been identified and improved, not only will cost savings be realized, but there will likely be a direct, positive impact on customer satisfaction and retention. Satisfaction increases lead to greater loyalty and repurchases, which lead to greater revenue, lower costs, and significantly improved profitability.

CASE 9: EDUCATION—USING COQ IN A UNIVERSITY ENVIRONMENT

Reference

Galati, Armand J. "COQ in a University Environment." *Quality Progress* 26, no. 10 (1993): 101–6.

In 1992, the University of Western Ontario (now Western University) established a quality center to drive quality and productivity improvement across the university. Their early efforts with process improvement teams were less successful than management had hoped. One of the major areas that had not been addressed was an effective means of measuring the opportunity for the teams and the savings that could be realized from the recommended improvements.

COQ has been used very effectively by manufacturing organizations for many years to identify opportunities and track improvements. With the service sector's increased interest in quality improvement, COQ has also been used as a measurement tool for its projects. COQ also had excellent potential in a university environment.

The manager of financial systems in the Department of Finance was convinced of the value of COQ as an effective tool for measuring quality improvement. He volunteered his department and identified two pilot projects: a major equipment replacement and a review of the telephone system. A project facilitator/trainer was assigned.

The primary concern was to ensure that COQ would be viewed as a *management tool* rather than a *financial tool*. As a management tool, COQ would enable process-improvement teams to manage their improvement efforts; as a financial tool, it could be seen as just another way to monitor employees.

For the pilot projects to be successful, the approach and methodology needed to be seen as equally important as the COQ measurement tool (Figure 8.14). Team members received extensive training in a disciplined approach to process improvement. An integrated process improvement and potential problem-solving model was used, supported by appropriate quality tools. In addition to the COQ, these included brainstorming, process mapping, and the seven basic quality tools: Pareto diagram, check sheet, cause-and-effect diagram, scatter diagram, histogram, control chart, and run chart.

Another key factor in the success of the pilot projects was management's support and guidance for the improvement teams. A key challenge for management was allocating time for team members to focus on improvement projects. Team members spent 25 percent of their time on their projects, meaning daily work was often delayed or offloaded. The teams needed to see quality improvement as part of their job, not something to be added to their duties. The key role of management was to monitor progress by focusing on the activities or steps of the team's process and noting the results of team efforts. By ensuring the team followed all the appropriate steps, the results would follow.

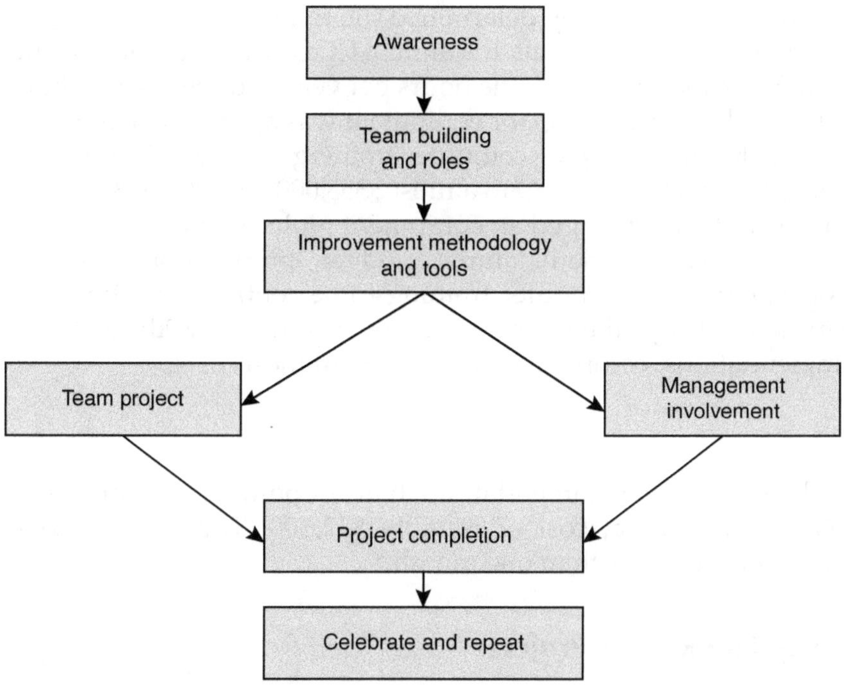

Figure 8.14 University process improvement team methodology.

Major Equipment Replacement Project

The purpose of this first project was to develop a process for equipment replacement within the finance department. Equipment included personal computers, copiers, facsimile machines, microfiche readers, and microfilm readers. The project would address individual needs, assessment, training, and communication. The team consisted of six members from the Finance Department.

The team initially documented and flowcharted the current equipment replacement process. They found that the decision to replace equipment was made by a single individual, using an informal process based on the best assessment of need. The team then conducted customer focus group meetings to gather information on perceptions of the current process. They found a lack of communication with staff regarding the current equipment replacement process and the absence of a feedback mechanism to notify individuals when a request for equipment was not approved. The perception was that the current method was not objective. In many cases, input from the individuals requesting the equipment was not taken into account.

The team then surveyed a sample of Finance Department members to determine the levels of equipment and training satisfaction. Using the collected information, the team estimated the COQ for the existing process. They based an estimate of productivity loss on the level of dissatisfaction expressed by the surveyed individuals. They determined the number of individuals using each type of equipment. As a result, the annual COQ was calculated by multiplying the number of individuals by the hours per year and the average hourly salary rate of $28. The annual COQ for equipment was approximately $210,000. They also calculated the COQ for computer training to be approximately $145,000, bringing the total yearly COQ to almost $355,000. (See Table 8.15.)

The team then developed an equipment replacement process using a team approach. This included annual budget preparation, repairs, reserve appropriations, and computer training to be centralized within the Finance Department. They then presented management with a list of seven recommendations, complete with implementation costs.

Result

In Table 8.16, the recommendations had a one-time cost of approximately $2,200 and an annual cost of roughly $12,500. The savings resulted in a payback period of less than one month.

Telephone System Review Project

The purpose of this second project was to review the existing telephone system and recommend improvements to maximize customer satisfaction at a cost-effective level. Again, six members of the Finance Department volunteered to participate in this project.

The team developed a telephone system review checklist to improve the number and types of calls being made to the department. Next, the team conducted a customer focus group to identify the problems that individuals contacting the department experienced. The most frequent issues identified were as follows:

1. Transfer of calls to the wrong area
2. Incomplete messages received
3. Phone pickup issues
    a. Picking up the same phone
    b. Picking up the same phone with a disconnect
4. Lack of notification when leaving the office
5. Expert not available when immediate help is needed
6. Inexperienced or temporary help answering calls
7. General calls are routed to the director of finance
8. Available options unused
9. Clearing of the voicemail box by the secretary

Table 8.15 Finance Department COQ for equipment and training.

Description	Total replies (a)	Weighted avg. of satisfaction (10 = max) (a)	Weighted avg. of dissatisfaction (b)	Est. productivity loss (b)	Finance staff (c)	Est. % of staff (d)	Weekly avg. hours used (e)	Weeks/ year	Avg. hourly salary (f)	Est. annual COQ
Personal computer	17	7.3	2.7	27%	45	89%	10	46	$28	$140,324
Copiers	19	7.1	2.9	O29%	45	100%	2	46	$28	$34,166
Fax	17	5.4	4.6	46%	45	89%	1	46	$28	$3,661
Microfiche	10	5.8	4.2	42%	45	53%	0.3	46	$28	$3,644
Microfilm	10	4.0	6.0	60%	45	53%	0.4	46	$28	$7,321
Estimated COQ for equipment										$209,316
Training (g)	19	7.5	2.5	25%	45	100%	10	46	$28	$144,900
Estimated COQ for equipment and training										$354,216

Notes:

(a) These schedules are based on the survey of nineteen members in the department.

(b) A satisfaction level of 10 (100 percent) indicated full productivity. A satisfaction level of 7 means 70 percent productivity and a corresponding 30 percent loss.

(c) Approximate number of nonmanagerial staff in the Department of Finance.

(d) Estimated percentage of staff involved is the number of responses divided by the sample size (e.g., 17/19 = 89 percent).

(e) Average weekly hours that equipment is used were determined by the survey.

(f) Average hourly salary for nonmanagerial staff is estimated at $28.

(g) When asked, only two people surveyed said they felt they had been adequately trained to perform computer-related functions. On a weighted-average basis, survey respondents reported that they were 75 percent trained to use their computers. Accordingly, lack of training is estimated to reduce personal computer productivity usage by 25 percent.

Next, implementation costs were estimated (Table 8.16).

Table 8.16 Estimated implementation costs.

Start-up costs		
Initial assessment of inventory and training	50 hours @ $28 =	$1,400
Database design	14 hours @ $33 =	$462
Database approval	4 people @ 2 hours @ $44 =	$352
Annual costs		
Annual time for team meetings	12 meetings @ 3 hours @ 3 people @ $33 =	$3,564
Ongoing support time for team meetings	12 meetings @ 3 hours @ 3 people @ $33 =	$3,564
Maintenance of inventory	1 hour/month @ $33 =	$396
Communications time	7 hours / month @ 1 person @ $33 =	$2,772
Estimated annual costs		$10,296
Total estimated costs for implementation: $12,510		

The team then summarized the COQ for the current process (Table 8.17). They first determined the time required to correct each of the first seven problem types. For example, the time wasted by calls transferred to the wrong area was 6.5 minutes. For the last two problems, paying for options not used and clearing phone mail, they calculated the cost of each. Clearing phone mail varied across departments, depending on the number of involved parties and the number of messages. Based on the information collected using the checklist, they calculated the COQ for each problem type. These varied in cost from area to area, depending on the frequency of errors. The total COQ for the telephone systems review project was approximately $65,500 per year.

The team then developed an employee telephone survey to solicit suggestions for improvement. Based on customer input, the team developed recommendations across eleven areas. The COQ was recalculated in accordance with these recommendations (Table 8.18).

Table 8.17 Finance Department COQ—before telephone system review.

Issue	Total cost
Transfer call to wrong area—6.5 min.	$10,991.99
Message not complete—6.5 min.	$638.38
Picking up the same phone—6.5 min.	$10,483.69
Picking up the same phone with a disconnect—8 min.	$4,692.00
Not telling someone when leaving office—6.5 min., 10 min., 13 min.	$9,090.75
Person needs immediate help, expert not there—15 min.	$9,237.38
Inexperienced or temporary help—5 min.	$306.67
General calls going to the director—10 min.	$1,564.00
Paying for options not activated or used	$660.24
Clearing the voicemail box	$10,190.44
Total	$65,477.02

Table 8.18 Finance Department COQ—after telephone system review.

Issue	Total cost
Transfer call to wrong area—0 min.	$0.00
Message not complete—0.5 min.	$48.88
Picking up the same phone—0 min.	$0.00
Picking up the same phone with a disconnect—0 min.	$0.00
Not telling someone when leaving office—2.5 min.	$5,131.88
Person needs immediate help, expert not there—2.5 min.	$1,539.56
Inexperienced or temporary help—0 min.	$0.00
General calls going to the director—0 min.	$0.00
Paying for options not activated or used	$0.00
Clearing the voicemail box	$10,557.00
Total	$18,852.81

Result

The COQ after implementation of the recommendations would be about $18,900 per year. Part of the new COQ included approximately $1,600 in training costs. This project resulted in annual savings of about $46,600.

Summary

These two pilot projects clearly demonstrated that the COQ is an excellent management tool. It has been shown that COQ can be used effectively to identify opportunities and track improvements not only in manufacturing and service sectors but also in the university community.

CONCLUSION

Across these cases, the lesson repeats. Define COQ in finance, select the denominators that leaders trust, and identify the few drivers that matter. Implement with a small, visible scope; then scale. Control gains with owners, thresholds, and simple reviews so that rework does not recur. Improve by reinvesting a portion of the savings into prevention, supplier capability, and team skills. Do this, and COQ becomes both the language of money and the logic of quality. It gives you a way to decide, to sustain, and to create value in factories, clinics, offices, code bases, and classrooms alike.

Chapter 9

Quality Systems

No COQ program stands alone. It's always part of a larger quality management system (QMS), which is woven into the fabric of the entire organization. You can't simply copy a quality system from one company and drop it into another—it won't work. Why? Because every organization operates under different conditions:

- *Strategy and environment:* Your business goals and external pressures shape your quality priorities.
- *Resources and risk:* Available people, time, and budget determine what's practical.
- *Culture and maturity:* History, work habits, and technology readiness influence how change takes hold.
- *Industry requirements:* Automotive, healthcare, and aerospace all demand different approaches.

Standards like ISO 9001 are written as broad requirements. They set the intent, not the method, leaving you to design a system that fits your context. Many industries build on ISO standards, adding requirements and issuing guidance documents, such as the ISO 10000 series, which provide practical implementation guidance. Other standards—environmental and calibration—may also apply, but they are beyond the scope of this chapter. Figure 9.1 illustrates how ISO 9001 is supported by ISO 10014 and ASQ TR 2-2024, and how the International Automotive Task Force (IATF) standard is an accredited standard to ISO, with requirements such as International Traffic in Arms Regulations (ITAR), Food and Drug Administration (FDA), and Sarbanes-Oxley Act of 2002 (SOX) making a small but nonnegligible influence on how systems meet the quality systems requirements.

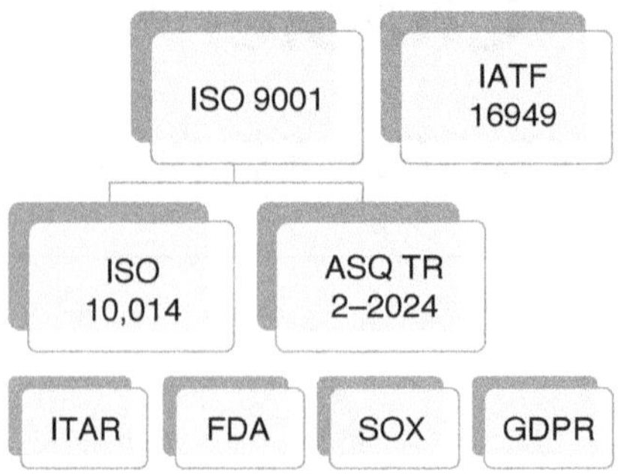

Figure 9.1 Relationship of systems and requirements.

ISO 9001: THE FLAGSHIP QUALITY SYSTEM AND COQ

We can start with ISO 9001 and examine how COQ metrics relate to each clause. ISO 9001 has ten clauses. We can skip the first three sections ("Scope," "Normative References," and "Terms and Definitions") as they are introductory and do not contain requirements.

Clause 4 is the "Context of the Organization," and 4.1 leads with, "The organization shall determine external and internal issues that are relevant to its purpose and its strategic direction and that affect its ability to achieve the intended result(s) of its quality management system." In the context of COQ this statement would require that, when an organization considers COQ, it begins with strategic direction and the identification of the most influential internal and external issues. This concept underscores the need for a custom COQ system.

Section 4.4.1(c) states, "The organization shall determine the processes needed for the quality management system and their application throughout the organization, and shall: c) determine and apply the criteria and methods (including monitoring, measurements and related performance indicators) needed to ensure the effective operation and control of these processes." While the ISO 9001 standard does not explicitly require use of COQ, COQ measurement of waste and monitoring expenses can clearly be included. In this same section, under (h), it states: "Improve the processes and the quality management system."

Clause 5 is "Leadership," and in addition to general statements in 5.1.1 about "ensuring that the resources" and "promoting continual improvement," there is more in section 5.1.2 on customer focus, "ensuring that: . . . c) the focus on enhancing customer satisfaction is maintained." Customer focus, along with 5.3(d), "ensuring the promotion of customer focus throughout the organization," is a strong suggestion to measure downstream consequences.

In clause 8, "Operation," 8.2.1(c) addresses customer communication: "Communication with customers shall include: c) obtaining customer feedback relating to products and services, including customer complaints;"

The clause does not state "cost of downstream consequences," but it is difficult to see how good decisions on customer feedback may be weighed without some consideration of the cost of such downstream consequences.

In clause 9, "Performance Evaluation," it says, "The organization shall evaluate the performance and effectiveness of the quality management system." Some may interpret this as the percentage of good products shipped to customers, but one of management's key roles is to ensure the wise application of resources. Without knowledge of waste-related expenditures and downstream consequences, along with monitoring expenses, top management will be unaware of how effectively the quality system performs. Indeed, under 9.3.2, "Management Review Inputs," it states, "The management review shall include . . . the adequacy of resources."

Finally, in 10, "Improvement" and specifically 10.1, "Continual Improvement," the standard states, "the organization shall continually improve the suitability, adequacy and effectiveness of the quality management system."

To fully and properly implement ISO 9001 and maximize its value, management needs sound cost data to monitor process-related expenses, waste-related expenses, and expenses arising from downstream consequences. Additionally, knowledge of investments to improve processes is required for management to make sound, forward-looking decisions.

Note: Quotations are taken from DIS/ISO 9001:2025, the draft of the sixth edition. As of the writing, expectations are that this draft will be very similar to the final release.

IATF 16949: WHEN INDUSTRY ADDS EXPECTATIONS (OR STUMBLING BLOCKS)

If you work in the automotive industry—or support it—you'll encounter an additional requirement beyond ISO 9001. The IATF standard 16949 adds explicit guidance: Management review inputs must include the COPQ, with text explicitly identifying COPQ as the cost of internal and external nonconformity.

Most industry-specific standards reinforce the same principles discussed under ISO 9001, even if they don't mandate a complete COQ metric set. IATF 16949 stands out for its widespread enforcement, which helps explain why many organizations adopt COQ, or COPQ, programs.

One nuance worth noting: the phrase *cost of poor quality* persists in many organizations, even though it reflects an older, narrower view. A mature COQ approach extends beyond COPQ to include prevention and appraisal costs. Unfortunately, auditors focused on wording can misinterpret this and flag a COQ program as noncompliant simply because it doesn't use the language "COPQ." That conclusion is incorrect. A well-designed COQ framework that includes COPQ is not only compliant; it signals a more robust, forward-thinking approach.

Key Takeaways

- In the management review, clearly show how your COQ data satisfies the COPQ requirement.
- If needed, add a note explaining that your COQ framework includes COPQ but extends beyond it for better decision-making.
- In the case of pedantic auditors, map your COQ categories to IATF's terminology (internal and external nonconfomity).

A LITTLE HELP FROM ASQ TR 2–2024: PRACTICAL GUIDANCE FOR COQ PROGRAMS

As an accredited standards developer, ASQ developed this document for optional use in conjunction with American National Standards. This document applies the ISO framework to improve alignment with other international standards, including clauses 1, "Scope;" 2, "Normative References;" and 3, "Terms and Definitions."

The definitions in this document include the following alternate terms for COQ: *prevention* is also listed as *investment*; *appraisal*, as *monitoring*; *internal failure cost*, as *waste*; and *external failure*, as *downstream consequences*.

Clause 4, "Context of the Organization," examines the needs of interested parties and notes that portions of an organization's operations may fall outside the scope of a COQ metric. The need to document the COQ process and its required subprocesses is discussed, with a recommendation to create flowcharts and Supplier-Input-Process-Output-Customer (SIPOC) charts.

Clause 5, "Leadership," emphasizes the importance of ensuring alignment with strategy, focusing on customer needs, and communicating responsibilities. Authorities related to COQ activities are assigned, communicated, and understood via policy, and appropriate costs are included in the COQ measure.

The TR document continues with implementation planning, resource and personnel assignment, risk-based thinking to enhance COQ metrics, and regular review of COQ metrics to ensure they meet current needs. Also in clause 7, knowledge, competence, and awareness are emphasized. Clause 8, "Operation," has a section on supplier COQ, along with suggestions for setting up a pilot COQ program. In Annex A, there is a decision tree for assigning cost elements into COQ categories. Annex B is a guide to starting a COQ program for small businesses.

Managing COQ Under Regulatory Pressure

In most industries, regulatory requirements do not significantly restrict COQ tracking, but some standards may affect how quality cost data is collected, stored, or shared. Organizations need to be aware of relevant regulations to ensure compliance, as in the following examples:

- ITAR (International Traffic in Arms Regulations) may restrict what quality cost data can be shared with third parties (or, in many cases, internally) or stored in cloud-based systems.
- ISO 9001 and other standards provide guidelines for documented information that influence how COQ data is maintained.
- FDA (21 CFR Part 820) requires documented quality records and controls, impacting how COQ data is collected, stored, and made available for audits in more specific requirements than ISO 9001.
- SOX (Sarbanes-Oxley Act) can influence the financial reporting of quality costs for publicly traded companies in the US.
- European Union's General Data Protection Regulation (GDPR) / data privacy regulations may restrict how quality data containing personal information of individuals in the European Union is stored or shared across borders.

Specific industry standards may also require minor adjustments:

- IATF 16949 (Automotive Industry Standard) mandates that suppliers implement a COPQ system, which is narrower than a complete COQ approach. Some auditors may challenge compliance if an organization uses COQ terminology instead of COPQ.

Again, in practice, a complete COQ system is more robust than a COPQ framework. However, if an auditor insists on strictly adhering to IATF 16949 terminology, organizations may need to clarify their approach or make minor adjustments to avoid unnecessary disputes. While adjusting COQ terminology solely for audit compliance is not recommended, minor wording modifications may be a pragmatic solution in some cases.

USING ISO 10014 TO FURTHER TRANSLATE QUALITY INTO VALUE

The ISO 10014 guidance document for realizing financial and economic benefits is aimed at leadership and outlines a top-down, structured, two-stage approach. Not intended just for quality, the procedures encompass environmental and occupational health and safety systems. It uses the quality management principles of

- customer focus,
- leadership,

- engagement of people,
- process approach,
- improvement,
- evidence-based decision making, and
- relationship management.

The economic benefits are realized by following these seven principles and applying them in a plan-do-check-act continual improvement cycle. A process approach and risk-based thinking, as defined in ISO 9001, are also guides to realizing these economic benefits. The same methods are used to sustain improvements.

Stages

In stage 1, top management is responsible for identifying and developing metrics to focus on desired improvements. In stage 2, these improvement goals are communicated to managers across the organization for implementation.

The standard includes four informational annexes:

- Annex A: Financial and economic benefits, related metrics, and linkages to the above quality management principles
- Annex B: Examples of a structured approach
- Annex C: Self-assessment tool
- Annex D: Application of quality management requirements

Stage 1

The self-assessment tool provides a baseline for implementation effectiveness. Gaps identified in the self-assessment may be addressed in line with the organization's strategic plans. A set of performance metrics is selected, validated, and reviewed to identify trends and risks. Such tools as dashboards, benchmarking, time series analysis, and sharing best practices (where applicable) are used to identify opportunities for action. Stage 1 concludes with assigning action plans to responsible individuals.

Stage 2

At this stage, process analysis is used within the plan-do-check-act cycle to advance action plans. At the end of each cycle, top management needs to review the results and determine whether they are acceptable.

Annexes in the ISO 10014 standard

Annex A contains example metrics that top leadership may use to determine the organization's status. While COQ is not explicitly mentioned here, components would serve as metrics to assess the state of the QMS.

Annex B provides three examples of both the stage 1 and stage 2 applications:

- generic application,
- small business application, and
- services application.

Annex C uses a five-level maturity assessment, with some questions for top management and others for process owners. The assessment has eleven questions.

Annex D shows which ISO 9001 clauses the steps in this standard apply to.

DIGITAL QUALITY SYSTEMS CAN HELP COQ, BUT ONLY SO MUCH

There has been growth in software options since 2007 to help organizations manage their quality systems. These software packages, often referred to as eQMS, can help reduce the overhead of managing a quality system.

Little is known in the literature about how various software packages interact with COQ metrics. So original research was conducted on May 5, 2025, at the ASQ World Conference on Quality and Improvement in Denver, Colorado, to provide a closer look at the current state of eQMS offerings. Sixteen vendors were interviewed, including both those that identified as eQMS providers and others whose systems managed one or more related functions, such as CAPA, documentation, or complaint handling. Document control–only systems were excluded from this analysis.

A strong influence from the life sciences sector was evident. Eight of the vendors focused on industries with heavy regulatory oversight, such as pharmaceuticals, medical devices, and food, reflecting requirements from ISO 13485, the FDA, and similar international regulations. One vendor specialized in automotive, another in general manufacturing, while the remaining suppliers served more niche or cross-industry applications.

When discussing COQ, many vendors showed limited familiarity with the term or with the legacy prevention, appraisal, and failure (PAF) model. Only one vendor reported actively developing a dedicated COQ module. Six vendors offered partial cost-tracking features, such as capturing defect costs via master cost lists or Enterprise Resource Planning (ERP) integrations, enabling standard costs to be tallied for issues like CAPAs. Seven vendors described how their platforms could be customized with low-code or no-code tools to capture certain quality costs, but these solutions did not provide a complete or standardized COQ framework.

A comparison between life science–focused systems and general manufacturing or automotive systems revealed differences in how costs are addressed. Fewer than 30 percent of life science–oriented systems included even basic cost capture, while approximately half of the general manufacturing and automotive systems supported rudimentary tracking. This suggests that while regulatory requirements drive functionality in life sciences, they do not necessarily lead to strong COQ integration. In contrast, sectors where cost efficiency is more central appear more likely to include cost-related features.

Overall, the study indicates that COQ remains underdeveloped in all eQMS platforms. Software providers appear eager to meet diverse customer needs, often relying on customization and flexible reporting tools to ensure their systems can adapt. However, cost-tracking features are only moderately present and far from ubiquitous. Organizations seeking robust COQ measurement will continue to depend on tailored solutions rather than fully integrated modules.

The publication "Advanced QMS For Dummies, 3rd ETQ Special Edition" (Crotty 2018) lists some ways a robust eQMS package can act to improve the COQ:

1. Reducing poor quality costs by decreasing scrap, rework, warranty, and recall expenses.
2. Preventing issues: Using QMS software to implement quality plans and processes proactively can help prevent quality issues.
3. Improving efficiency and reducing waste: A QMS can help identify problems early in the production process, preventing costly delays.
4. Tracking and measuring: Trending nonconformances and other quality events helps identify systemic issues that drive the COPQ.
5. Investments that aid prevention and deliver ROI: QMS investments can yield significant returns, including reduced scrap and rework costs.
6. Predictive analytics: More advanced QMS implementations with predictive quality analytics can identify potential issues before they occur, helping to lower the costs of poor quality.

Care should be taken in selecting a software vendor. Consider what features are offered, and map these to organizational needs; determine the training needs for initial installation and ongoing training needs, what support will cost, and what it will cover, especially as upgrades occur; hardware and network needs to run the software; assess how long the vendor has been operating and how many updates the software has seen; decide how many other systems need to be linked or integrated with the QMS software; determine what has been the experience of other firms who purchased the software and, of course, the initial price of the software.

If your organization's QMS is in flux, remember that people are much more flexible than automated packages. You may need to wait to implement software until you think you have a relatively stable management system.

CONCLUSION

A quality system only works when it fits the organization that has to live with it. COQ belongs inside that system, not beside it, and it earns its keep when leadership uses it to make decisions about resources, risk, and customer impact. Standards like ISO 9001 give the intent, but you have to supply the design, the discipline, and the follow-through. Do that well, and COQ stops being a report. It becomes a management habit that keeps the QMS effective, credible, and funded.

KEY TAKEAWAYS

- *Design the QMS for your context* (strategy, risk, resources, culture, industry), not for a textbook or another company.
- *Tie COQ to ISO 9001 clauses you already manage*, so COQ supports process control, performance evaluation, and continual improvement.
- *Make COQ a standing input to management review* with trends, top cost drivers, and clear actions and owners.
- *Use a simple COPQ-to-COQ mapping for IATF 16949* to prevent audit debates from becoming a time sink.
- *Set data rules early* for access, storage, retention, and sharing when regulations apply (FDA, ITAR, SOX, privacy).
- *Document the COQ process so people can run it* with clear definitions, sources, estimate rules, and a review cadence.
- *Treat eQMS as workflow support, not a COQ solution*, and automate only after the system is stable enough to deserve software.

Chapter 10

Leading with COQ: From Metrics to Management

Up to this chapter, we have studied COQ from nearly every angle. We have defined COQ, built tools around it, presented assorted case studies, and applied it across products, services, and software. What remains is the connecting layer that decides whether any of it works: people. Projects get chosen, funded, and sustained by people. They bring judgment, bias, creativity, and leadership. The practical job in this chapter is to earn buy-in for the right projects and to link COQ to the choices that shape execution, strategy, and innovation.

This chapter shows how to work with people in the language of COQ. Starting with "Lead with the Numbers"—putting ROI and payback in the box executives trust—we then move to "The Human Side of COQ," where numbers align around bias and narrative. Next, "Using COQ as a Management System" shows how leadership runs with COQ—using governance rhythms and a scorecard to make strategy visible and turn metrics into clear commitments teams can deliver. "Quality in the Wider System: Finance, ESG, and Industry 4.0" connects those choices to markets, stewardship, and digital signals, so trade-offs are priced and risks managed. "COQ Maturity Model" explains how to build capability and sustain gains. We finish with "Trends and Quality: The Leadership Imperative"—what's next and how to lead it.

LEAD WITH THE NUMBERS

Put the Number in the Box

Imagine this: A quality manager walks into the boardroom with a detailed project plan on defect reduction, cycle time improvement, and rising customer satisfaction. The charts are crisp, the results are real, but the room is quiet. A finance director flips to the last page, squints at the numbers, and then the CEO politely asks, "So…what's the payback?"

That scene might sound hypothetical, but for me (Pruitt), it was all too real.

Early in my career, I stood, a young engineer, at the front of a conference room, rehearsed and ready. A PowerPoint deck glowed behind me. I'd poured weeks into this project. The data was solid. The analysis revealed a clear opportunity. We had assembled an interdisciplinary team to investigate, measure, analyze, improve, and control. It was DMAIC, start to finish, by the book.

But I lost the audience.

I can't recall if the CEO looked at his phone or just shifted in his seat, but I knew I'd missed the mark. He was the kind of leader who listened closely. Yet the story I told didn't match the one he needed to hear.

Afterward, a trusted mentor summed it up perfectly: "You didn't put the number in the box."

That phrase stuck with me. I hadn't led with the punch line: the cost savings, the payback, the financial impact. I buried the lede. And no matter how compelling the data or how disciplined the analysis, I hadn't framed the message around what mattered to him: the profit and loss, the 10-K (a comprehensive annual report that every publicly traded company in the United States is required by law to file), the balance sheet, and the statement of cash flows—the documents investors, regulators, and board members use to measure success.

It was a missed opportunity but a valuable lesson. One I've carried with me into every project, COQ or otherwise, since.

How to Use This Chapter

This chapter is not a deep dive into finance or accounting but a practical guide, using all the previous chapters, to help quality professionals engage executives, make stronger business cases, and frame work in financial terms that drive decision-making.

Along the way, we'll identify why quality initiatives often meet resistance. The reasons are not limited to competing priorities. Cognitive biases such as the sunk cost fallacy and confirmation bias also play a role. By recognizing these factors, you will be better equipped to design strategies that account for both the rational and emotional sides of decision-making.

We will also explore how such tools as executive dashboards and balanced scorecards can integrate quality metrics that align with strategic goals. In addition, environmental, social, and governance (ESG) considerations are becoming more aligned with well-managed quality systems, offering yet another pathway to connect quality work with broader business priorities.

Ultimately, the goal is clear: to communicate the value of quality in ways that matter to the people who decide what gets funded, maintained, or scaled, and what gets shut down. Doing the work is not enough. To drive change, we must connect that work to the outcomes in which leaders are held accountable.

When that inevitable question comes—"What's the payback?"—you will know which number belongs in the box and, more importantly, why it matters.

What Leaders Are Accountable For

The earlier story wasn't just about a missed number; it was about a mismatch in framing. The failure was twofold: not speaking the language of money and not presenting the story in a way that aligned with the listener's priorities.

To overcome this kind of resistance, the conversation must begin with financial impact. Using money addresses the first fold and, starting with it, the second. And no one better illustrates this perspective than Milton Friedman.

In his 1970 essay for *The New York Times*, "The Social Responsibility of Business Is to Increase Its Profits," Friedman famously wrote: "There is one and only one social responsibility of business: to use its resources and engage in activities designed to increase its profits." Friedman's model positions the firm as a fiduciary mechanism: Managers act on behalf of owners and are responsible for maximizing shareholder returns. For decades, this view shaped executive priorities and boardroom discussions. Profitability (ethically and within the "rules of the game") wasn't just a goal; it was the primary obligation.

But this model has faced growing criticism. In recent years, the definition of value creation has expanded. Businesses are being asked to consider not just shareholders but stakeholders: customers, employees, suppliers, communities, and the environment. ESG frameworks, purpose-driven strategies, and sustainability initiatives all reflect this shift.

COQ, while a long-standing tool, fits neatly into both paradigms. On the one hand, it speaks directly to profitability by reducing waste, rework, and failure costs. On the other hand, it supports long-term resilience and stakeholder satisfaction by building more robust and reliable systems. Whether your company reports in dollars, yuan, euros, yen, pounds, rupees, or Swiss francs, COQ turns quality into a language every CFO, and CEO, understands.

And that's the key: Quality doesn't exist in a vacuum. It must be framed in a way that resonates with both traditional business thinkers and those championing modern corporate responsibility. COQ provides that bridge.

The full spectrum of business storytelling or narrative frameworks is not appropriate to cover here; nor is there room since it is a broad, nuanced field. Instead, this chapter keeps things practical. A few simple adjustments in how you frame quality initiatives can reinforce credibility, clarity, and relevance.

The core insight is this: The CEO in that boardroom wasn't a perfectly rational decision-making machine. They were human. That's not a flaw; it's a strength. But it does mean we need to be deliberate when communicating financial value, especially as we lean into the economics of quality.

The Human Side of COQ Decisions: Why Numbers Alone Don't Convince

In that earlier presentation story, what was missing wasn't data; there was a ton of data, but it was dollars and objective, financial impact. But over time, experience reveals that even in finance, numbers don't always carry the day on their own.

Consider the stock market. Company names come and go, but there are always a few trading at valuations that far exceed what their fundamentals (cash flow, earnings, and assets) can justify. Why? Because they have a story. A narrative that captures imagination and persuades people to believe in something bigger than the balance sheet.

The same dynamic plays out in boardrooms, office meetings, and on factory floors. Credibility, persuasion, and narrative all shape how ideas gain traction. Marginal projects can succeed with the proper framing. Complex decisions can be justified even when the math looks uncertain, because people believe in the broader vision.

This isn't about manipulating the truth. It's about recognizing that humans make decisions, not spreadsheets. And leaders respond to stories that make sense of complexity, especially when results are still emerging, or the numbers don't tell the whole story.

To drive quality through COQ, professionals need both the financial grounding and the narrative skill to explain why it matters. We build credibility, influence strategy, and move from measurement to meaningful change by managing biases, emphasizing value, creating empathy, acknowledging cultural aspects, and communicating effectively.

Biases That Bend Decisions

Humans are not purely rational actors; our decisions are shaped as much by psychology as by logic. This is especially true in finance and operations, where cognitive biases often distort judgment. Quality professionals are already familiar with how statistics can mislead when tools are misapplied, and finance has its own traps. A prime example is the sunk cost fallacy—continuing to invest in a failing initiative because resources have already been committed. The instinct is to "throw good money after bad," even though sound financial tools like ROI focus on future value rather than past losses. This is compounded by loss aversion, where the pain of losing feels greater than the benefit of equivalent gains.

Other biases—such as confirmation bias, overconfidence, short-term thinking, and groupthink—can cloud financial judgment. Leaders are not immune. They may resist new cost-transparency initiatives, arguing that if the company is profitable, there is no need to dig deeper. Quarterly pressures and a false sense of security can lead them to overlook hidden costs in scrap, rework, warranty claims, and lost opportunities.

For these reasons, introducing a COQ program requires more than executive approval; it demands alignment in understanding. A well-prepared presentation can counter resistance by revealing the gap between current waste and potential gains. Highlighting both the behavioral tendencies that drive poor decisions and the tangible financial impact of hidden costs creates a compelling case for change. To speak the language of finance effectively, quality leaders must master not only the numbers but also the human behaviors that influence how those numbers are interpreted.

Position Quality as Value Creation

The way COQ is framed matters. Instead of emphasizing compliance or control, present it as a tool for value creation, improving profitability, reducing waste, and strengthening both customer trust and employee engagement. When quality is positioned as a driver of growth rather than a cost to contain, it captures executive attention. Leaders want performance gains but often hesitate over how to achieve them without overspending or disrupting operations. A clearly presented COQ program offers a practical, confidence-building answer.

Build the Case with Numbers and Empathy

Build empathy with narrative. A strong business case balances data with narrative. Lay out what the program entails, what data will be collected, and the improvements to expect. Numbers make the case tangible, but examples make it persuasive. For instance:

> For every $100 in product sales, we are losing $10 to poor quality, specifically rework. Plan A is projected to reduce rework cost by 50%. This is priority because currently we sell the product for $110 and a competitor sells for $107.

Typically, a cost-savings argument cites only the dollar figure or perhaps the product's profit margin. But the case for reducing rework by 50 percent is not just about direct savings. It may also create a competitive advantage by strengthening the company's ability to increase sales volume and gain market share. Depending on the audience, that implication may be inferred, made explicit, or left out altogether. In another instance, it could be stated "the savings will be over one million per year" or "We could save the cost of a new Corolla every week."

Either of these statements can be highly effective, provided they are factual, as they must be. Each delivers emotional impact in a slightly different way. Large round numbers, such as one million dollars, often create a sense of urgency. Yet a comparison to something tangible and familiar can make the same value feel more concrete. The two examples may be similar in monetary terms, but consider which is easier to visualize: a number on a page or fifty-two Toyota Corollas parked in the parking lot?

The appropriate and effective persuasion approach will be unique to your audience. The idea is not to exaggerate but to make it personal, something that the audience can relate to.

Beyond facts, persuasion also requires empathy. Leaders are people under pressure, wary of committing to initiatives that might fail. Resistance is often less about opposition than about uncertainty—position executives as the ones who recognize the opportunity and act decisively. By aligning the narrative with their perspective, you give them ownership—and ownership fuels commitment.

From Approval to Action: Overcoming Cultural Resistance

Winning executive approval is only the start. Cultural resistance often lingers, rooted in taboos about waste, skepticism from past failed efforts, or fatigue with change. Address this with open conversation: Assess where resistance might surface, invite honest feedback, and create space for real dialogue. Cultural nuances may be part of differences in nationality, geography, size, history, and organizational structures. Regardless, when barriers are named, solutions emerge.

To succeed, COQ must be everyone's business, from design to procurement, operations to logistics. Position the program as a shared responsibility connected to strategic goals. Finally, end presentations with clarity, not just inspiration: Outline next steps, expected outcomes, and how ROI will be measured. Visible sponsorship and cross-functional engagement transform COQ from analysis into action.

Communicating COQ in All Directions

At this point, COQ, as a tool to control and improve, has been discussed at great length, but it has at least one other power: communication when leaders use it as a shared language. *Upward,* frame it in the terms boards already watch. Tie prevention investments to cash, margin, capacity, and risk. Put one number in the box and a short story around it. "This investment cuts downstream consequences by X dollars over the next four quarters, reduces complaint exposure in our top segments, and is protected by these control measures." Use a simple pair of signals, so directors can track progress. For example, "If we are investing in design-review tollgates and selecting a supplier based on Parts per Million (PPM) performance the first quarter, we can project the followed lower warranty cost per unit starting in the third quarter."

Downward, treat COQ as cultural signposts. People want to know what good looks like. Celebrate first-time right and clean handoffs—post visible trend lines for waste and downstream consequences at the team level. Use fast feedback that connects today's action to next week's results. When a line fixes a chronic rework loop, highlight the capacity gained and the customer callbacks avoided. People will know where the organization's priorities lie by what gets discussed, measured, and acted upon.

Across, take the same categories to suppliers and partners. Make external failure a joint issue with shared triggers and response times. Fold COQ into vendor scorecards and contracts. Reward prevention wins on both sides. A small library of standards that clarifies containment, escape response, and audit closure can save months of friction. Do the same with channel partners for service escapes.

Throughout, communicate choices plainly. When budgets tighten, say what you will stop and what risks will increase. When you invest in prevention, say what you expect to avoid and how you will verify it. That candor is paramount. It builds trust, strengthens leadership credibility, and supports a mature quality culture so that COQ, or any system, can be used effectively by leaders.

Using COQ as a Management System

COQ cuts waste and errors and, used well, can become a management lens that links daily work to long-term goals. It exposes hidden inefficiencies, guides investment in products, services, and systems, and earns executive attention when translated into ROI, margin, and cash flow. Framed in finance,

COQ stops being a quality metric and becomes a business tool that shapes strategy, improves profitability, and lifts enterprise performance.

Table Stakes: Scrap, Rework, Warranty

An organization that isn't at least tracking scrap and rework is operating without visibility to key issues. Whether you call it COQ or not, those costs are tangible, measurable, and actionable. Ignoring them is the equivalent of flying blind.

Some companies have gone far beyond the basics, building sophisticated COQ programs that continuously generate returns. But the truth is, even the most advanced systems start with the fundamentals. Scrap, rework, and other failure and waste costs are the table stakes of operational insight. Without them, there's no baseline from which to improve.

The phrase *economics of quality* has led to confusion over the actual business value of quality management. Some argue that the phrase is misleading, because poor quality is never economical. Others take the opposite view, claiming it's too expensive to aim for 100 percent quality and that "good enough" is sufficient. Both of these positions miss the point.

The real challenge arises when leaders within the same organization hold conflicting views. One executive pushes for near perfection, while another tolerates a certain level of defect or inefficiency. If that misalignment goes unchecked, quality efforts will always fall short of their full strategic value.

A strong COQ program brings clarity and alignment. But to get there, we must draw a line between quality costs and the cost of the quality organization. This distinction is often misunderstood. Quality costs are not the salaries of the quality team or the QA Department's operating budget. They are the expenses incurred when work is not done right the first time (or incurred to prevent the costs).

The most visible examples include:

- reworking defective products,
- retesting completed assemblies,
- rebuilding broken tools, and
- correcting transaction errors or customer records.

Others are less visible but just as real:

- repurchasing materials due to supplier failures,
- redesigning faulty components,
- handling customer complaints and returns, and
- reprocessing a loan application or food order in a service setting.

In short, any cost that would not exist if everything were done correctly the first time adds to your COQ.

Nearly every function in a business can contribute to these costs, either through action or inaction. That is the essence of operational waste. Ignoring it is naive. But settling for only identifying and reducing waste is also limiting. COQ processes and initiatives can and should do more.

They can support strategic improvements. They can align departments around shared goals. And they can transform quality from a passive cost center into an active driver of performance, innovation, and profitability.

How COQ Drives Business Outcomes

One of the most critical roles of the quality function is to help leadership see how long-term quality performance shapes both profitability and reputation. At its best, quality is not a supporting act; it is a strategic function, equal in importance to finance, operations, marketing, and R&D. Strategic planning for quality must be deliberate, not reactive.

Without building quality into operations from the start, organizations risk delivering short-term gains at the expense of long-term satisfaction. From the first design concept through final delivery and support, quality must align with customer needs throughout the product or service's life cycle. That journey often spans months or years.

Customer satisfaction cannot be left to chance. It must be actively planned, measured, and continuously improved. This is where the voice of the customer (VOC) becomes central. VOC methods, such as surveys, interviews, complaint tracking, and net promoter scores, help translate customer needs into actionable requirements. These tools enable distinguishing between basic expectations, performance features, and potential "delighters." Delighters are unexpected features or experiences that exceed expectations and leave a lasting impression. When designed thoughtfully, they create loyalty and differentiation, even in competitive markets.

Integrating VOC data into the COQ framework directly links operational performance to customer outcomes:

- High levels of rework or returns may signal a failure to meet basic expectations.
- Warranty claims may reveal gaps in performance features that matter most to customers.
- Investment in prevention activities, such as improved design reviews or training, can directly address customer pain points uncovered through VOC.

This is how quality efforts shift from internal compliance to external value creation.

Over time, organizations have adopted various improvement methods to address different types of challenges, including:

- balanced scorecards,
- Enterprise resource planning (ERP) and customer relationship management (CRM) systems,
- process improvement steering committees,
- custom dashboards and reporting platforms, and
- enterprise-wide initiatives, such as lean and Six Sigma.

These systems help identify areas for improvement, track performance, and align teams. But without financial measurement, many of these efforts struggle to secure sustained executive support. COQ fills that gap by providing a clear financial linkage.

By turning improvement from a local initiative into a strategic tool that resonates with leadership, COQ data quantifies the costs of defects, errors, inspections, and quality investments. It gives departments the language to include quality-related costs in their planning and budgeting processes. More importantly, it positions quality as a value driver, not just a support function.

When integrated into the strategic planning process, COQ creates a path for quality to influence high-level decisions. It enables the quality team to speak directly to executive concerns: profitability, risk, customer retention, and long-term growth.

The result is a system that not only drives operational efficiency but also deepens customer satisfaction and strengthens business performance. And it does so in a way that invites participation across the organization. Everyone, from product design to frontline service, has a role to play in reducing the COPQ and increasing the value delivered to customers.

Sidebar Intel 1994: Prevention Deferred, Brand Pays

Intel's Pentium division's flaw started as a technical issue and became a board-level decision. The company's first instinct was to limit replacements. IBM halted shipments of Pentium-based PCs on December 13, 1994, amid concerns about the defect. Intel subsequently took a $475 million pretax charge related to the replacement program. Through a COQ lens, the path is clear. Deferred prevention and incomplete containment shift costs to external failure, lost capacity, and reputational risk. The price is set publicly, not in the spreadsheet.

Keep Wins from Slipping

Improvement gets the spotlight. Control holds everything together. One draws attention and celebration. The other keeps the gains from slipping away.

Control is what protects the investment in quality. It ensures that progress made today becomes the standard tomorrow, not just a passing high point. And much like cost avoidance, its actual value is often in what never happens:

- the defects that don't return,
- the costs that don't rise, and
- the customer complaints that never appear.

In many organizations, quality control spending follows a predictable cycle. When problems surface, resources are added. When those problems seem resolved, resources are cut. This reactive pattern creates a stop-start effect: quick wins followed by gradual losses. Improvement and control should work together, not trade places.

Some reductions in oversight may be appropriate once stability is achieved. But the key is not how much you reduce; it is what you keep. COQ helps teams monitor this balance. It provides visibility into where risks are building and where investment is still needed.

This is where intelligent monitoring comes in. By tracking quality costs over time, teams can

- identify early warning signs before they escalate,
- avoid significant, sudden failures,
- reinforce processes that are performing well, and
- target new control efforts where variation is creeping in.

Ask yourself:

- Do we know what stable performance looks like in each process?
- Are we solving the same problems again and again?
- Are our metrics showing absolute control or just temporary gains?

If the answers are uncertain, it may be time to refocus on control. A robust COQ system supports both sides of the equation; it helps you solve problems and keep them solved.

When it comes to strategic planning, quality deserves a seat at the table not just because it delivers improvements but because it preserves them. And that kind of consistency, year after year, is what turns operational performance into a strategic advantage.

Sidebar Alcoa, 1987–2000: Safety Drives Performance

Paul O'Neill chose worker safety as the leading indicator for running Alcoa. He required immediate incident transparency, fixed hazards at the source, and treated safety as the daily test of process capability. Within a year, profits hit a record. By the time of his retirement, annual income was roughly five times higher, and the market value had risen by about $27 billion. Call it COQ, or cost of safety, without the label. It is the same arc many quality leaders pursue. Invest in prevention, build control, and watch external failures fall along with their costs.

Quality as a Competitive Lever

A central mission of the quality function is to help leadership understand the long-term impact of quality performance, not only on operational efficiency but also on the company's profitability, reputation, and market position. These are not theoretical concerns. They influence every key decision an organization makes.

To support this mission, quality must be embedded into the strategic planning process alongside finance, operations, marketing, and R&D. Unless quality is deliberately designed into every stage, from early concept through delivery and long-term customer use, an organization cannot be confident it is meeting expectations or sustaining competitive advantage.

Over time, various programs have been adopted to manage performance and pursue improvement. These include balanced scorecards, ERP and CRM systems, digital modeling and digital twins, steering committees, and enterprise-wide initiatives like lean and Six Sigma. Each offers tools for tracking, diagnosing, and acting on performance data. What most lack is a direct financial measurement of quality. Filling the gap is where COQ delivers unique value.

COQ provides the financial lens that makes enterprise performance measurable in terms that matter to executives. It translates improvement into cost impact, allowing top leadership to compare quality initiatives directly with other investments. And unlike one-time business cases, COQ supports ongoing alignment with corporate goals.

This alignment spans every layer of the organization:

- customers, whose satisfaction is linked to lower failure costs and better design;
- employees, who engage more deeply when their work avoids waste and drives value;
- managers and executives, who can make smarter decisions when quality data is financially grounded;
- board members and shareholders, who expect sustainable growth and capital efficiency; and
- vendors and partners, who are more accountable when quality expectations are measurable and tracked.

COQ becomes a unifying framework, a way to connect daily operational choices with long-term business objectives.

It also helps address a frequent and damaging issue: short-term thinking in financial decision-making. While finance teams are essential for maintaining fiscal discipline, there can be a tendency to prioritize immediate savings over long-term value. In this environment, quality programs, primarily preventive or strategic initiatives, are often mislabeled as discretionary costs.

By highlighting how investments in quality reduce downstream consequences and recurring waste, COQ offers a precise counterbalance. It reframes quality spending as an investment rather than an expense. And more importantly, it helps leaders resist the urge to cut programs that do not show immediate payback but are essential to sustained performance.

Action is where COQ becomes more than a reporting tool. It becomes a communication tool, one that helps good leaders make better decisions, even under pressure. It reinforces the idea that short-term gains are not worth long-term setbacks and that cutting investments in preventive quality measures today often guarantees consequences and waste tomorrow.

Like ESG frameworks and balanced scorecards, each of which is discussed further, COQ brings discipline and visibility to long-term planning. It ensures that quality is not only managed but also protected and prioritized to sustain the organization's success.

When COQ data backs quality, it is no longer just a technical or compliance topic; it is a clear link to the bottom line. It becomes a key part of business strategy. It earns its place in resource planning, capital allocation, and executive conversations. When quality is elevated to the C-suite, it transforms from a department function into a source of competitive advantage.

Put COQ in the Room

COQ creates value and is a competitive advantage when it shapes decisions. That means moving it out of a quality spreadsheet and into the conversations where capital is allocated, risks are weighed, and priorities are set. Start with the board agenda. Tie COQ to the same outcome language directors already use: growth, margin, cash, risk, and reputation. When a proposal reaches the board, include a one-page COQ view that answers three simple questions:

1. What quality costs will this decision avoid or create over the next twelve to twenty-four months?
2. How will it affect customer trust and the likelihood of complaints?
3. What control measures keep gains in place after launch?

Put the number in the box; then show how it is protected.

Avoid the trap of treating COQ as an isolated administrative task. If it lives only inside the quality function, it becomes a report, not a lever. Co-own the numbers with finance and operations. Define categories, data sources, and calculation rules together to ensure a single version of the truth. Build COQ into monthly business reviews, the annual plan, and the capital stage gate. Add it to commercial processes too. New product charters, supplier contracts, and service-level agreements should include explicit COQ targets and escalation triggers. If incentives ignore COQ, behavior will, too. Align bonuses and objectives so that fewer downstream consequences and lower waste are rewarded alongside revenue and throughput.

Alignment with organizational key performance indicators (KPIs) is where COQ earns its keep. Map each COQ measure to a KPI that leaders currently track. External failures and downstream consequence costs are linked to customer retention, warranty rates, and brand health. Internal failure and waste costs sit under cost to serve, overall equipment effectiveness, and first-time right; investment and monitoring link to capability growth, audit closure, and process stability. Set owners, targets, and review cadences. Show leading and lagging pairs, so people can see cause and effect. For example, an uptick in design review yield and supplier PPM should forecast a drop in claims and returns next quarter. Keep the math visible and straightforward. A small set of stable, repeatable measures beats a long catalog nobody reads.

Framing COQ as risk mitigation is equally essential. Boards think in terms of risk appetite, compliance exposure, and brand trust. Translate quality failures into that language. A spike in waste costs might be an early warning of downstream consequences and reputation loss. Reworking that choke capacity is a supply risk. Missed calibrations and incomplete records are compliance risks. Place the top COQ drivers on the enterprise risk heat map with likelihood, impact, and current controls. Include near-miss counts, not only realized losses, since they reveal where a headline event could originate. Tie remediation plans to specific controls, such as mistake-proofing, supplier containment, or training, and show the expected reduction in risk-adjusted cost.

Short-term pressure will always test discipline. COQ helps leaders resist cutting prevention investments and monitoring for a quick cosmetic saving. Frame prevention investments as working capital protection and brand insurance. For example, "Cutting $500,000 from training and incoming inspection saves cash this quarter but increases expected warranty by 1.2 million over the next two cycles." Make the trade-off explicit. Decisions may still favor near-term constraints, yet everyone will understand the consequences and can plan accordingly. COQ is practical governance, not accounting theater.

Operationally, integrate COQ with the scorecard, so it sits beside financial, customer, process, and learning indicators. That balance keeps liquidity and profitability in view while preserving attention on capability and satisfaction for the long term. ESG often addresses some of these concerns in its way, and COQ usually supports those aims by reducing waste, improving safety, and strengthening supplier quality. The point is alignment. Customers, employees, managers, executives, board members, stockholders, and vendors should see the same story, in the exact numbers, at the same time.

Sidebar Xerox, late 1980s to early 1990s: Finance and Quality Share One Playbook

Xerox's U.S. Marketing Group made COQ a management tool rather than a report. MIT Sloan documented how the division defined categories, embedded the measures into ops cadence, and used COQ to guide decisions. Management accounting summaries credit the program with sizable first-year savings and continued gains. At the same time, the company's broader quality push won the 1989 Malcolm Baldrige Award and helped reverse a slide in world market share. The lesson for leaders is simple. When finance co-owns COQ, and it appears in plan, review, and incentives, prevention spending survives budget pressure and pays back through lower downstream consequences.

Scorecard: Make Strategy Visible

Quality, as a discipline, emerged in part to fill gaps left by traditional financial metrics. Income statements and balance sheets show what happened but not why. They may point to declining margins or rising warranty costs, yet they don't explain the process failures, poor training, or broken feedback loops that caused them.

In much the same way, the balanced scorecard (BSC) was created to move leaders beyond short-term financial results toward a more complete, strategic view of performance. Both COQ and the scorecard are tools that ask more profound questions: What do we value? And how do we measure what matters?

First introduced by Robert Kaplan and David Norton in their 1992 *Harvard Business Review* article "The Balanced Scorecard—Measures That Drive Performance," the framework proposed four categories for organizing performance data:

- financial,
- customer,
- internal business processes, and
- learning and growth.

The scorecard was never just a dashboard. It was designed to translate strategy into measurable outcomes, to link vision with execution. Within only a few years, adoption had spread across industries. By the early 2000s, Bain & Company survey data showed that it was used in more than one-third of large organizations. Its flexibility and adaptability across business models, industries, and maturity levels made it highly scalable. Tools followed. Research followed. Recognition followed.

But over time, something else followed too, a superficial "scorecard mentality."

In some organizations, the scorecard became a template to fill out rather than a tool for shaping strategy. The original intent, to align action with purpose, was lost in translation. Metrics were chosen without strategic grounding. Data was displayed but not used. The scorecard became a decoration instead of a direction.

The same risk exists with COQ. Collecting defect data or tallying rework costs is not the goal. The value lies in linking those metrics to strategic outcomes. Used well, the scorecard and COQ are not just measurement systems. They are frameworks for alignment. They guide attention, shape dialogue, and drive decisions.

Gaining Balance from the Balanced Scorecard

Though the original BSC includes four perspectives, organizations often tailor the framework to fit their unique needs, just as COQ programs should be designed to reflect the business. The point is balance. A good scorecard includes the following:

- *Short-term financial metrics,* such as cash flow, are needed to manage liquidity and operational risk. Liquidity metrics protect the company from near-term collapse, even when long-term performance looks strong.
- *Profitability and performance metrics,* including cost drivers and improvement KPIs, are used to monitor the factors that most directly affect financial outcomes.

- *One or two long-term metrics* should focus on sustainability, resilience, or capability. Metrics could include modern ESG factors or, as initially intended, learning and development. It might reflect investments in R&D, employee capabilities, or COQ indicators that show customers are well served now and into the future.

This structure helps organizations make better trade-offs. It doesn't eliminate tension between short-term and long-term goals. But it creates a language to talk about them and a system for understanding consequences.

Map COQ to the Scorecard

COQ is one of the few quality tools that fits across all four perspectives. It enables a more complete understanding of how quality impacts performance, not only operationally but also strategically.

- **Financial:** COQ provides direct measures of the cost of poor quality, return on prevention investment, warranty spend per unit, and the total impact of failures on gross margin and profitability.
- **Customer:** External failures and downstream consequences can serve as early indicators of customer dissatisfaction or lost trust. Complaint trends, resolution costs, and net promoter scores can all be integrated alongside quality performance.
- **Internal business processes:** COQ data offers process-focused metrics, scrap, rework, first-pass yield, correction cycles, and inspection effort that link directly to lean and Six Sigma initiatives.
- **Learning and growth:** Often overlooked, this perspective is where quality training, structured problem-solving, audit follow-through, and employee-led improvement efforts live.

Organizations that invest in these areas often see stronger long-term COQ trends.

Some organizations have renamed and reframed COQ categories under other strategic themes. Consciously, or organically coming to similar conclusions, they come up with a compelling set of categories that made immediate sense to executives:

- protect customer loyalty (downstream consequences),
- drive first-time success (waste),
- invest in capability (investment), and
- strengthen confidence (monitoring).

This subtle shift in language helped reposition COQ not as a cost-tracking system but as a tool for delivering on strategic commitments.

Protect Long-Term Value

In today's business environment, short-termism remains a challenge, especially in finance. Even profitable companies can fail if cash flow isn't managed. Investments in prevention programs often get cut when they don't show an immediate return. Long-term initiatives are shelved in favor of quarterly performance.

COQ, embedded into a scorecard, helps correct this imbalance. It provides data that links quality investments to avoided costs, reduced risk, and improved customer retention. It gives leaders the language to explain why some investments are worth keeping, because they prevent much larger future losses.

And like ESG indicators, COQ can help shift the conversation from short-term optics to long-term health. It enables more intelligent trade-offs and more transparent decision-making.

Align the Organization

Most importantly, integrating COQ into the scorecard supports alignment. When quality data appears alongside financial and customer metrics:

- Executives begin to view quality as a strategic driver.
- Managers and frontline teams see how their work contributes to outcomes that matter.
- Employees understand that reducing waste and defects isn't just about compliance; it's about delivering value.
- Boards and shareholders gain confidence that investments in quality are being measured, monitored, and aligned with business goals.
- Even vendors and partners can be engaged using these same measures, creating a shared understanding of expectations.

When scorecards reflect COQ data, quality becomes everyone's job, and everyone understands why it matters.

Connect Metrics to Moves

Neither the scorecard nor COQ can eliminate complexity. But together, they simplify communication. They clarify expectations. And they help organizations operate with consistency and purpose.

This combination doesn't just measure performance. It connects it to strategy. It helps turn insight into alignment, and alignment into impact.

Sidebar Quanterix, 2022–23: Making Quality Economics Visible to Investors

Facing an assay remediation effort, Quanterix changed cost allocation to reflect quality activity in the cost of goods and began reporting pro forma margins to "give greater visibility" into the quality work. Management told investors to expect near-term margin pressure and to watch sequential improvement as remediation reduced rework and scrap. That is COQ logic used in public: Put prevention on the income statement, make the trade-offs explicit, and let the street track progress.

TOOLBOX: WHAT TO USE, WHEN

Executive Checklist

COQ has the most impact when leaders use it to guide choices. The goal is simple. Turn quality data into decisions about where to invest, what to stop, and how to grow. The checklist below helps frame executive COQ reviews, followed by a guide to connect familiar tools to leadership use. The intent is opportunity, not penalty.

Checklist of executive reflection questions for COQ reviews:

- Are the top COQ drivers clear, ranked, and tied to owners and dates?
- What value is released if we cut waste and downstream consequences by 25–50 percent?
- Which metrics are leading, and which are lagging? Do we see control or only short bursts of improvement?
- How do COQ trends affect cash, margin, capacity, and customer retention this quarter and next year?
- Where are we accepting short-term gains that create higher future costs?
- What risks to reputation, compliance, and brand trust are visible in downstream consequences and external failure costs, including near misses?
- Do supplier and partner scorecards reflect the same COQ categories and targets?
- Is the data trustworthy? Have we confirmed the measurement system and definitions with finance and operations?
- How will we fund investments in prevention and capability without increasing total cost?
- What is the plan to keep the gains? Who owns the control plan, and what triggers a management review?

Role of Classic Quality Tools in COQ

The checklist turns COQ into a boardroom asset. Now classic quality tools can be used to answer executive questions, not just to complete projects.

Pareto: Rank the few causes that drive most COQ. Use for capital allocation and quarterly focus.

FMEA: Convert quality risks into a living entry on the enterprise risk register. Tie actions to risk reduction and expected COQ impact.

Lean and Value Stream Mapping: Show where waste inflates costs. Quantify capacity unlocked and its effect on revenue and lead time.

Six Sigma and SPC: Link variability to yield, rework, and margin loss. Use capability targets to justify process investment.

DOE: Test high-leverage changes fast. Present expected COQ reduction with confidence intervals, not anecdotes.

VOC, QFD, Kano: Connect design choices to customer satisfaction, delighters, and lifetime value. Tie the external failure and downstream consequence cost trends to these insights.

A3 and 8D: Create executive-ready narratives—one page, from problem to countermeasure and to verified effect on COQ.

MSA and Gage R&R: Validate that the numbers used to guide decisions are reliable. If the measurement is weak, fix it before funding.

Poka-yoke and standard work: Frame these as risk controls that prevent recurrence. Quantify avoided costs and compliance protection.

SIPOC and process ownership: Clarify boundaries and handoffs so accountabilities for COQ categories are unambiguous.

Control plans: Show how improvements will be sustained, who watches which signals, and what triggers action.

Lead with the Top Line

Lead with the upside. COQ reductions are not only savings. They are offering pricing flexibility, faster delivery, higher throughput, better cash conversion, and fewer customer defections. Ask the following:

- What growth becomes possible if we free 10 percent of capacity by eliminating rework?
- What price moves could we make if warranty costs drop by half?
- Which contracts become viable if supplier defects fall below the target?

Present prevention and monitoring as protection of the margin and brand, not as overhead. Make trade-offs explicit so that short-term cuts do not quietly create long-term losses.

QUALITY IN THE WIDER SYSTEM: FINANCE, ESG, AND INDUSTRY 4.0

Financialization, ESG, and Stewardship

Financialization may sound unusual in a quality context, yet it matters here. As capital markets grew more dominant, feedback loops shortened, and quarterly optics began to shape everyday choices. Profit is vital, but the emphasis on near-term signals often crowds out investments that create durable value. Quality as a management discipline arose, in part, to correct that imbalance. During industrialization, feedback loops were disrupted by the growth of organizations and the division of labor. Today, we encounter the same challenges, exacerbated by relentless, finicky market expectations. COQ gives leaders a way to weigh today's expense against tomorrow's waste, service escapes, and brand damage. It makes the long term visible in numbers that finance respects.

COQ acts as a counterbalance to short-term pressure. Prevention and monitoring become investments with clear payback rather than nice-to-have programs. When leaders can point to avoided external failure costs, preserved capacity, and steadier cash conversion, they have the evidence needed to resist cuts that look attractive this quarter but erode performance next year. Resistance is important in governance because it prevents the organization from trading small savings now for significant losses later.

Quality investment also connects directly to ESG. The link is practical. Lower scrap and rework reduce energy use and the amount of materials sent to landfill. Safer processes protect people and communities. Strong traceability, audits, and supplier controls improve compliance and board oversight. When COQ metrics sit alongside ESG indicators on executive dashboards, leaders can see how a design review, a supplier fix, or a training cycle reduces both costs and environmental burdens while strengthening trust.

Stewardship is the leadership lens that ties profit to responsibility. Directors and executives have duties that include care, loyalty, and good faith. COQ helps fulfill those duties by translating operational choices into risk, return, and reputation outcomes. A simple practice works well. For each major decision, include a one-page COQ view: expected change in failure cost over the next four quarters, effect on margin and cash, customer impact, and the control plan that keeps gains in place. The conversation shifts from "What does this cost?" to "What value do we protect and create?"

There is also an ethical side to quality costs. Prevention protects lives, livelihoods, and the environment, not just margins. A product that works the first time avoids customer harm and field waste. A safe process avoids injury and the ripple effects on families and communities. COQ gives leaders the language to treat prevention as both good business and the right thing to do. That clarity builds social trust, which in turn supports growth, regulatory confidence, and talent retention.

In practice, bring this perspective into the systems you already run. Embed COQ alongside financial, customer, process, and learning indicators in the scorecard. Tie the most significant external failure drivers to the enterprise risk register. Align supplier scorecards to the same categories, so partners see the cost and ESG impact of escapes in real time. Most importantly, openly narrate the trade-offs. When you must cut, state the risk you accept and how you will watch for trouble. When you invest, state what you expect to avoid and how you will verify results.

This stewardship approach prepares the ground for maturity models in the next section, COQ maturity. COQ will serve as the spine of that model, showing how organizations progress from ad hoc accounting to integrated, values-driven management in which profit and responsibility move hand in hand.

Digital Integration: COQ in Industry 4.0

COQ also changes how digital systems are designed, what they display, and how people act on them. When COQ metrics are embedded in dashboards and enterprise platforms, the system shifts from reporting activity to reporting impact. ERP, manufacturing execution systems (MES), CRM, and business intelligence (BI) tools share a single COQ data model with shared definitions, owners, and refresh cycles. That single view makes portfolio reviews faster, capital cases more precise, and monthly business reviews more consistent. The critical shift is governance. Data stewardship follows the money, so quality, finance, and operations align on one truth.

Predictive analytics and AI get sharper when COQ defines the target. Instead of flagging generic anomalies, models rank risks based on projected failure costs. Industry 4.0 signals from machines, suppliers, and field service are converted into dollar exposure by product, line, and customer segment. That turns alerts into priorities. A spike in scrap becomes an expected cash impact, with recommended actions and a simple confidence band. Leaders can decide earlier, with clearer stakes.

Digital COQ also changes how you work with partners. Supplier portals and shared scorecards use the same well-defined COQ categories, thresholds, and triggers. Vendors see the cost impact of escapes and the value of prevention investments in your language, not just theirs. Joint workflows open a case; when variation creeps in, assign owners on both sides, and close only when downstream consequences fall. The critical outcome is trust. Everyone sees the exact numbers at the same time.

Most of all, COQ pushes holistic systems thinking. It links design choices to service cost, supplier performance to brand risk, and training to yield. Trade-offs are no longer framed as isolated metrics. They are shown as end-to-end cost and value. This perspective keeps short-term cuts from eroding long-term health and helps the organization learn. When COQ sits within your digital stack, the enterprise operates more like a single system, guided by a clear understanding of what quality is worth.

COQ MATURITY MODEL: CAPABILITY NOW, DIRECTION NEXT

The scorecard provides direction for controlling and improving your core process; a COQ maturity model is a map to enhance your COQ process. It aligns leaders on where they are, where they are going next, and the few moves that matter. Think in four levels. Reactive tracks visible failures and audits in pockets. Proactive measures track prevention investments, monitoring, and waste with basic, primarily executive reporting. Progressive links costs across functions and uses results to guide decisions, suppliers, and culture. Predictive covers the full spectrum, shares insights widely, and applies multiyear analytics to steer investment. Use the table and statements to select your current state and plan improvement.

The ASQE/IOE 2025 survey asked leaders to place themselves on a similar scale, and the distribution is informative. About 12 percent identified as reactive, 25 percent as proactive, a majority (52 percent) as progressive, and 11 percent as predictive. Treat that as a benchmark and an encouragement. Many organizations live in the middle, and the next steps are visible.

Use this simple COQ maturity model (Table 10.1) as a map to align leaders on where you are today, where you want to be next, and the few moves that matter. Keep the language plain. Let the table carry the detail and let the four self-assessment statements anchor the discussion.

Table 10.1 COQ maturity model.

Category	Level 1—reactive	Level 2—proactive	Level 3—progressive	Level 4—predictive
Completeness	Limited to inspection and defect costs (internal and external)	COQ across monitoring, waste, and escapes, with some accounting of investment	Investment, monitoring, waste, and escapes are linked to iterative feedback systems (such as internal audit) programs	Full spectrum, including advanced downstream consequences
Scope	Limited to core operations; reactive and fragmented	Applied in multiple functions or departments; partly integrated	Integrated across departments with aligned metrics and shared accountability	Extended enterprise-wide; goals cascade to every level and quality is embedded in the organization's DNA
Analytics and technology	Ad hoc spreadsheets	Templates and dashboards with historical trends	Integrated ERP/QMS with multi-year analysis	Advanced analytics, prediction, and simple scenario simulation
Reporting	Local and siloed reports	Results go to executives only	Results are visible to executives and the front-line employees	Results shared across all levels and with customers, suppliers, and other stakeholders
Outcomes	Explains failures after the fact	Supports compliance, efficiency, and complaint monitoring	Guides strategic improvements, supplier management, and cultural change	Informs investment choices; predictive prevention; external trust

Use the statements below for the quick self-assessment your multidisciplinary team will compare against during the review:

Level 1—reactive: Our organization has made some effort to measure the costs of poor quality, focusing mainly on COPQ and relying on inspections or audits. Reporting is local and fragmented, often in spreadsheets. Results are used reactively with minimal attention to prevention expense.

Level 2—proactive: We measure COQ across prevention investments, monitoring, and failure waste categories, though prevention is underdeveloped. Deployment is partially integrated but still centered on production, and reporting is limited to upper management. Some trends help with compliance and efficiency.

Level 3—progressive: We connect prevention investments, monitoring, and waste costs, including rudimentary elements of downstream consequences (i.e., escapes), and align quality cost metrics across multiple groups. Sharing results with leadership and frontline workers to drive strategic improvements and cultural change.

Level 4—predictive: We use permanent, multiyear COQ analytics to demonstrate the value of control and improvement. Downstream failure expense is tied to rapid internal changes. Goals are set at all levels, and we capture the full spectrum of investments, monitoring, internal failure waste, and downstream consequences.

Evaluation should be short and evidence based. The chart provides details for deeper analysis and shared understanding; the four statements are what the multidisciplinary team uses to compare their practices and pick a level. Bring a small group from quality, finance, and operations. Read the four statements aloud and select the single best fit for each category in the table. Ground the choice with a few artifacts: one COQ trend that links to margin or cash, one customer signal such as complaints or warranty per unit, one supplier metric with owners and dates, and one active control plan that keeps gains in place. Mark a twelve-month target and a three-year aspiration. If definitions do not align with finance or the measurement system is weak, fix that first so people trust the numbers.

Improvement works best one step at a time. From reactive to proactive, publish shared COQ definitions with finance, issue a one-page COQ report for your top product or service, add a basic customer lens for complaints and warranty, and stabilize two critical measures. From proactive to progressive, extend COQ to more functions with named owners and budget lines, pair leading and lagging indicators so cause and effect are visible, mirror your COQ categories in supplier scorecards, and run one joint containment with proof of closure, and share results in plain language to build pride in first-time right. From progressive to predictive, embed COQ fields in ERP, QMS, MES, CRM, and BI, so alerts include a dollar forecast; build multiyear analytics that quantify avoided downstream costs; pilot simple predictive models that rank actions by expected failure cost avoided; and consider a brief external view for customers or investors to build trust.

Keep a steady cadence. Use a monthly business review with a one-page COQ map, a quarterly strategy check to confirm investments are funded, and control plans are working, and establish clear thresholds that trigger action when signals drift. Protect prevention in the plan. When cuts are necessary, state the risk you accept and how you will watch for trouble. When you invest, state what you expect to avoid and how you will verify results. This discipline is essential for credibility and for reinforcing a positive culture.

Close by tying the model back to leadership. COQ is both the language of money and the logic of quality. It helps you persuade with honest math. Be sure to align metrics with strategy so teams pull together, with clear communication up, down, and across, so suppliers, partners, boards, and front lines see the same picture, and practice stewardship by treating prevention as value creation for customers, communities, and owners. Leaders who master COQ are not just cost managers. They shape culture, strategy, and future value. Pick your current level, set a target, choose a few actions that matter; review, learn, and repeat. Progress builds trust, and trust turns COQ from a report into a way of leading.

TRENDS AND QUALITY: THE LEADERSHIP IMPERATIVE

What's Next

"Change is the only constant" is a saying often attributed to Heraclitus of Ephesus. If life is changing, then so are organizations and technology. Change has always been constant, yet in our interconnected, technology-driven world, it feels faster every year. The next decade will be more connected and quicker. Smart devices will get smarter. Products will report their own health. Supply networks will move in real time. Data will be abundant; attention will be scarce. In that future, leaders need tools that turn raw signals into priorities. COQ is not the only tool, but among quality methods, it fits exceptionally well. COQ translates drift, scrap, service escapes, and warranty risk into dollars at stake. That is how you choose what to fix first, where to place capital, and how to explain tradeoffs to boards and teams.

Leaders around the world already see the road ahead. Results from the ASQE/IOE 2025 survey show that 85 percent of respondents agree they understand how quality costs affect financial performance. COQ data reaches the C-suite in 70 percent of organizations. Encouraging, yes, and there is room to improve. This book aims to help both quality professionals and executives, including those who wear both hats, close the gap. We focused on modern language, a firmer grasp of accounting and finance, and human communication. Paired with a firm understanding of COQ fundamentals to drive improvement and control, these practices set the stage for a future in which the financial framing of quality is routine in executive forums.

Maturity will not be uniform, and that matters for competitiveness. More than half of organizations (52 percent) report that they have connected prevention, monitoring, and failure costs across functions and are using the results to guide strategy and culture. A smaller frontier group, 11 percent, IoE terms "predictive," runs multiyear CoQ analytics, ties downstream consequences to rapid internal changes, and sets quality cost goals at every level. Those two signals mark the journey from local accounting to enterprise learning.

The same survey highlights what can hold you back. Common barriers include technology limitations on holistic analysis at 15 percent, lack of team cooperation at 11 percent, and ineffective reporting at 12 percent. The highest reported barrier for both quality professionals and executives was the lack of senior leaders' support for prioritization. Even with the perennial concern about leadership support, these are addressable with clear ownership, a shared COQ data model, and disciplined review cadences.

Global competition raises the stakes. Leaders report tracking downstream consequences that directly affect reputation, compliance, and loyalty. Brand or market value impact is measured at 59 percent, legal expenses at 50 percent, warranty at 45 percent, recalls at 41 percent, and customer service at 48 percent. In a world where one post can turn a local defect into a global event, measuring these categories is not optional.

Position COQ as part of a future-oriented leadership toolkit. Embed COQ categories into ERP, MES, CRM, and BI, so alerts carry a dollar forecast, not just a code. Use predictive analytics and industry 4.0 signals to rank actions by the expected failure cost they avoid. Tie machine data, supplier feeds, and field-service tickets to prevention choices you can explain in cash, capacity, and complaint exposure. Share the same frame with suppliers and partners so containment and prevention are a joint work. Map the most significant external-failure drivers to the enterprise risk register and to the scorecard, so boards can see how a design review or supplier fix moves risk, cost, and customer trust at the same time.

Treat COQ as a leadership tool for uncertainty and complexity. Markets will move. Networks will shift. Expectations will climb. COQ helps you see choices as systems rather than fragments. It links first-time right on a single line to margin, cash, and brand across regions. It is a daily discipline and strategic guide in one.

Quality's Call to Leadership

Here is the simple truth. COQ is both the language of money and the logic of quality. It gives you numbers that finance trusts and stories that people can follow. It turns customer loyalty, process stability, and risk into one narrative the enterprise can act on.

The task is straightforward. Persuade with honest math and plain talk. Align metrics with strategy, so teams pull in the same direction. Communicate up, down, and across, so suppliers, partners, boards, and front lines see the same picture. Practice stewardship by treating prevention as a value-creation opportunity for customers, communities, and owners.

Leaders who master COQ are not just managers of cost or quality. They are shapers of culture, strategy, and future value. They build systems that learn, investments that endure, and trust that carries the brand through hard times. That is important work.

So take the next step. Put COQ at the center of your decision-making. Use it to set priorities in your digital stack. Use it to frame risk in board conversations. Use it to encourage people to take pride in their first success. Lead with courage and foresight. The future will be instrumented and fast. Let COQ be your clarity, your compass, and your promise that the gains you make today become strength for years to come.

OUR FINAL OVERALL CONCLUSIONS

If this book has a single arc, it is simple: Quality is not a department, and COQ is not a report. COQ is how you connect planning, control, and improvement to the decisions leaders are actually held accountable for: cash, margin, risk, capacity, and customer trust. Put the number in the box, tell the truth about trade-offs, and build the habits that keep gains from slipping. That is how COQ stops being a metric and becomes a way of leading.

GLOBAL TAKEAWAYS: WHAT TO CARRY INTO MONDAY MORNING

- *Put the number in the box.* Lead with payback, ROI, and cash impact, then earn belief with a short story and a control plan. If you bury the number, you bury the project.
- *Juran first, PDCA always.* Plan the system, control the gains, improve the causes, repeat. Most teams fail by skipping the quiet middle step.
- *Definitions beat debates.* Choose investment, monitoring, waste, and downstream consequences, or other names. Reach an agreement with finance and operations, then stop arguing and start fixing.
- *Normalize the noise, or you will chase ghosts.* Use the right denominators (sales for long range, operational bases for short range), so trends reflect reality, not volume swings.
- *Table stakes are scrap, rework, and warranty.* If you are not tracking at least those, you are guessing. COQ starts with visibility, not ambition.
- *Control is steering, not paperwork.* Set baselines, drift triggers, and review cadence. Control plans are the price of keeping wins.
- *Make COQ cross-functional by design.* Quality costs are often caused upstream and paid downstream. Report costs to the process owner, not just the cost center.
- *Suppliers are part of your COQ, whether you admit it or not.* Hidden supplier costs live in your plant, your travel, your schedule, and your customer

complaints. Use Pareto analysis, scorecards, and joint CAPA to reduce total cost.

- *Software changes the timing, not the logic.* Agile spreads investment and monitoring into smaller cycles and pulls waste earlier. Measure lightly enough to keep speed, but consistently enough to learn.
- *Run COQ as a management system, not a side project.* Put COQ in the management review, the scorecard, the risk register, and the capital stage gate. If incentives ignore COQ, behavior will, too.
- *Maturity is a choice.* Pick your level, pick the next move, and execute for twelve months. Progress builds trust, and trust keeps prevention funded when budgets tighten.
- *Today, stewardship is an enterprise-level expression of quality.* COQ helps leaders tell the truth about short-term savings that lead to long-term losses, and it ties quality to ESG, safety, compliance, and brand resilience within a single, clear framework.

Appendix A

Financial and Accounting Concepts for Quality Management

Every organization must account for what it spends to create value. Managers, investors, and regulators want to know: What did it cost to deliver this product or service, and what profit remains after expenses?

At its core, profit equals revenue minus cost. Those costs are typically classified as direct or indirect. From there, they flow into intermediate calculations, such as cost of goods manufactured (COGM) and cost of goods sold (COGS) and ultimately to net income. These exact figures also feed into the standardized financial statements used by stakeholders to evaluate performance and guide decisions.

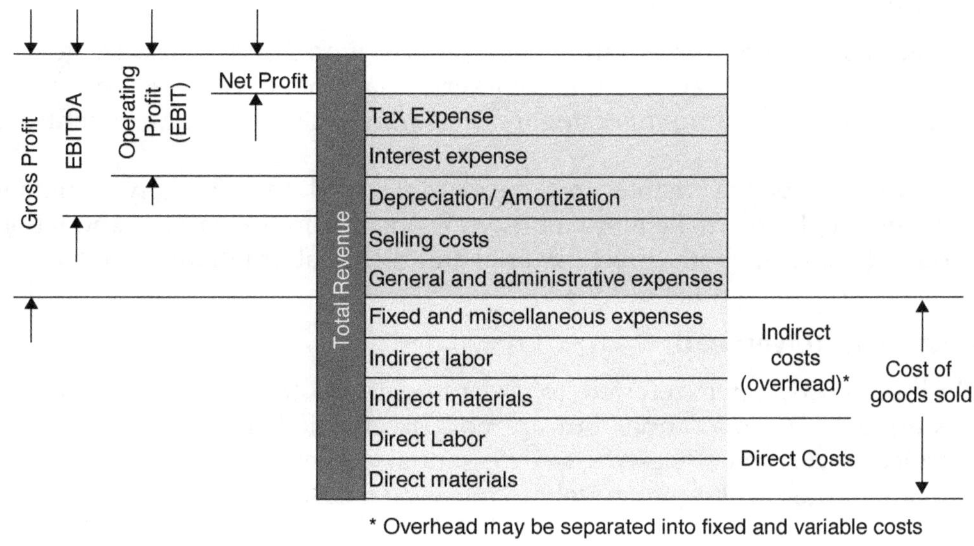

Figure A.1 Conventional cost and price structure.

This appendix provides a practical foundation for understanding cost structures, financial statements, and key profitability metrics. It then explores management accounting tools for budgeting, forecasting, and overhead allocation, followed by project evaluation methods like net present value (NPV), internal rate of return (IRR), and payback. We close by linking quality initiatives to enterprise value through valuation models and financial frameworks. Each section is designed to help quality professionals translate operational improvements into financial outcomes that resonate across the business.

Figure A.1 is a visual representation of the following section.

COST STRUCTURES

Direct Costs

Direct costs are those that are clearly and easily attributable to the production of a good or the delivery of a service.

Direct Materials

These are the raw materials, components, and supplies that go into the finished product. In a manufacturing plant, this includes steel, plastics, or electronics assemblies. In a hospital, it could consist of surgical instruments, syringes, or implants. Items like janitorial supplies or safety gloves are not included here because they do not become part of the product itself.

Direct Labor

Direct labor is the compensation of employees who directly produce goods or provide services. Factory line workers, machine operators, and service technicians all fall into this category. It usually includes wages, benefits, and payroll taxes.

Direct costs are visible and measurable, and they usually scale with production. If you make more units, you consume more material and labor. If you reduce scrap, both direct material and direct labor fall immediately.

Indirect Costs (Overhead)

Indirect costs, often referred to as *overhead,* are the supporting expenses required to run operations but are not directly tied to any single product, service, or customer. These costs are essential to keeping the business running but are harder to assign precisely to individual cost units.

Indirect Materials

These are items consumed in production but do not become part of the final product. Examples can include lubricants, cleaning agents, maintenance supplies, and small tools used in production. The distinction from direct materials lies in their supporting role rather than their presence in the end product.

Indirect Labor

This category includes employees who support the production process but do not physically build the product or deliver the service. Typical examples are production supervisors, quality inspectors, maintenance technicians, material handlers, warehouse staff, and plant administration.

Fixed and Miscellaneous Expenses

These include facility costs, such as rent, property insurance, utilities, equipment depreciation, and security. Some expenses, such as depreciation and rent, are generally fixed and remain constant regardless of production level. Others, such as utility costs or maintenance, may vary to some extent with activity.

Fixed vs. Variable Overhead

Unlike direct costs, which tend to vary directly with production volume, overhead behaves in more complex ways:

Fixed overhead stays the same in the short term, regardless of output—for example, facility rent, salaried supervisor pay, and depreciation.

Variable overhead increases or decreases with production activity—for example, indirect materials like machine lubricants, utility usage, and certain types of maintenance.

Step costs are a hybrid: They remain fixed over a specific range of activity but jump when a threshold is crossed. Hiring a second-shift supervisor when production expands is a typical example.

Understanding which costs are fixed and which are variable is consequential because it changes how costs behave as volume changes and, therefore, how decisions should be made. Fixed costs must be covered regardless of output, while variable costs scale with activity. This distinction becomes especially relevant in budgeting, forecasting, and break-even analysis.

COGM

COGM represents the total production cost of goods completed during a specific period. It includes both direct costs (materials and labor) and allocated indirect costs (overhead). To calculate COGM, you add up the costs incurred during the period and adjust for changes in work-in-process (WIP) inventory:

$$\text{COGM} = \text{direct materials} + \text{direct labor} + \text{manufacturing overhead} \\ + \text{opening WIP} - \text{closing WIP}$$

This formula reflects the cost of producing goods for sale but excludes distribution, marketing, and administrative expenses.

COGS

COGS builds on COGM by including any changes in finished goods inventory:

COGS = opening finished goods + COGM – closing finished goods

COGS reflects the total manufacturing cost of the products actually sold during the period. In manufacturing environments, it's tied closely to production activity. In service businesses, an equivalent measure is sometimes used to represent the direct cost of delivering services.

Beyond production, some interpretations include additional categories that support the delivery of the product or service to the customer. These are not part of formal COGS under accounting standards, but they are often included in internal analysis.

Selling Costs

These expenses relate to customer acquisition and product delivery: sales salaries, commissions, advertising, warehousing, order processing, shipping, and billing.

General and Administrative (G&A)

These broad operational support costs, such as finance, legal, human resources, IT, and other overhead, are not tied directly to sales or production.

Assume no change in WIP or finished goods inventory.

See Table A.1 for the following example of building a cost structure.

Worked Example: Building a Cost Structure

Imagine a company produces 10,000 units of a product.

Direct materials: \$25 per unit → \$250,000

Direct labor: \$15 per unit → \$150,000

Indirect materials: \$50,000 (total)

Indirect labor: \$100,000 (supervisors, inspectors)

Fixed expenses: \$200,000 (rent, depreciation, insurance)

Selling costs: \$120,000

G&A costs: \$80,000

Total costs:

Direct = \$400,000

Indirect/overhead = \$350,000

COGM = \$750,000

Total operating costs before tax = \$950,000

Table A.1 Cost structure.

Item	**Amount ($)**
Direct materials	250,000
Direct labor	150,000
Indirect materials	50,000
Indirect labor	100,000
Fixed expenses	200,000
Selling costs	120,000
G&A costs	80,000
Total direct	400,000
Total indirect	350,000
COGM	750,000
Total operating costs before tax	950,000
Revenue (10,000 × $120)	1,200,000
Profit before tax	250,000

If the company sells the 10,000 units for $120 each, revenue is $1,200,000.

Profit before tax = $1,200,000 – $950,000 = $250,000.

This layered build-up illustrates how costs accumulate and why managers must pay attention to both direct and indirect categories.

Together, direct costs, indirect (overhead) costs, and SG&A expenses account for the majority of line items on the consolidated financial statements. While only direct production costs are included in formal COGS for external reporting, understanding the complete cost picture, including selling, general, and administrative expenses, is indispensable for internal decision-making, pricing strategy, and profitability analysis.

COST CENTERS VS. PROFIT CENTERS

One common distinction in management accounting is between cost centers and profit centers. These terms are widely used, sometimes loosely or inconsistently. While the label itself may not always be applied rigorously, the real difference lies in how performance is measured. That distinction carries practical consequences, which is why it's worth clarifying.

Cost Center

A cost center is a department, function, or business unit whose performance is evaluated based on cost control rather than profit generation. These units provide internal services or operational support and are expected to operate efficiently within a set budget.

Examples include:

- HR,
- IT,
- maintenance departments, and
- quality assurance or compliance labs.

In some organizations whole manufacturing locations are considered cost centers. The manager of a cost center is responsible for managing expenses and supporting operations but does not directly influence revenue.

Profit Center

A profit center tracks both revenue and costs, allowing its profitability to be measured directly. These units operate more like standalone businesses within the company and are held accountable for their financial results.

Examples include:

- product lines,
- retail locations,
- regional sales offices, and
- service divisions.

A profit center manager is responsible for growing revenue, managing costs, and ultimately delivering profit.

Investment Center

An investment center is a step beyond a profit center. In this case, the manager is also responsible for capital investment decisions and is evaluated on return on assets or return on invested capital. This model is often used in divisional structures where leaders have broader control over strategy and long-term planning.

Why the Distinction Is Sometimes Blurry

In practice, the same organizational unit can be treated differently depending on context. For example, a manufacturing plant might be considered a cost center if its role is simply to produce parts efficiently, while sales and pricing are managed elsewhere. But in another organization, the same plant might be treated as a profit center if it runs as a business unit with control over output, pricing, and customer contracts.

In companies with strong front-end functions, such as sales, marketing, product management, and strategy, manufacturing is often viewed purely as a cost center, even though it plays a direct role in enabling sales. This flexibility in how units are classified can lead to confusion, but in most cases, the financial treatment matters more than the label. Still, it's helpful to understand how the distinction affects measurement and decision-making.

When the Distinction Matters

- *Performance evaluation*: A cost center is measured by efficiency and cost control. A profit center is evaluated based on revenue generation and profitability. Trying to apply the wrong metric leads to misleading conclusions.
- *Resource allocation*: Profit centers typically compete for capital investment by showing return potential. Cost centers usually compete for budget based on their efficiency and strategic role.
- *Strategic alignment*: In some cases, organizations turn cost centers into internal service businesses by using *chargebacks* or *transfer pricing*. For example, an internal IT department might be restructured into a shared services group that charges business units for support.

Key Takeaway

A cost center is evaluated on cost efficiency. A profit center is evaluated on profitability. This distinction shapes how managers view financial reports and how capital decisions are made.

FINANCIAL STATEMENTS

Every company, from a corner shop to a multinational corporation, must summarize its financial activity using a standard format. While complexity and sophistication scale, economic performance is communicated through three primary financial statements:

- Balance sheet: a snapshot of the company's assets, liabilities, and equity at a specific point in time.
- Profit and loss (income) statement: a report of revenues and expenses over a given period, showing whether the business earned a profit or incurred a loss.
- Statement of cash flows: a reconciliation of cash inflows and outflows across operations, investing, and financing activities.

Together, these statements provide insight into whether the business is stable, profitable, and liquid. For internal decision-making, they also underpin investment evaluation, performance monitoring, and financial planning.

Balance Sheet

The balance sheet answers a simple but powerful question: What does the company own, and how is it financed?

It is organized into three sections:

- Assets: resources the company owns or controls, such as cash, accounts receivable, inventory, property, and equipment.
- Liabilities: obligations the company owes to others, including loans, accounts payable, and accrued expenses.
- Equity: the owners' residual interest is the interest after liabilities are subtracted from assets; equity includes retained earnings and contributed capital.

A fundamental accounting equation governs the balance sheet:

- Assets = Liabilities + Equity

This equation must always hold because every dollar of assets is financed either by borrowing (liabilities) or by the owners (equity).

Worked Example: A Simplified Balance Sheet

Table A.2 A simplified balance sheet.

Assets	Amount ($)	Liabilities and equity	Amount ($)
Cash	200,000	Accounts payable	150,000
Accounts receivable	300,000	Bank loan	250,000
Inventory	250,000	**Total liabilities**	400,000
Equipment	400,000	Equity	750,000
Total assets	1,150,000	**Liabilities + equity**	1,150,000

In Table A.2, the simplified balance sheet shows a company with $1.15 million in assets. It owes $400,000 to creditors, and $750,000 belongs to the owners.

Why It Matters

Liquidity

Can the company meet its short-term obligations? A strong position in cash and receivables relative to accounts payable and other short-term liabilities indicates good liquidity.

Leverage

How much of the business is funded by debt versus equity? In this case, a $250,000 loan against $750,000 in equity suggests a relatively conservative capital structure.

Investment Capacity

A company with solid equity and moderate debt levels is typically better positioned to fund new projects, access financing, or absorb short-term shocks.

Profit and Loss (Income) Statement

While the balance sheet provides a snapshot at a point in time, the profit and loss (P&L) statement, also called the *income statement*, tells the story of performance over a period—typically a quarter or a year. It begins with revenue, subtracts the various costs of doing business, and ends with net profit. Table A.3 provides an example of a simplified income statement.

Typical Format

- Revenue (sales)
- COGS: includes direct costs and manufacturing overhead, adjusted for inventory changes
- Gross profit = Revenue – COGS
- Operating expenses (e.g., R&D, admin, marketing)
- Operating profit = Earnings before interest and taxes (EBIT)
- Interest and taxes
- Net income = Profit after tax

Worked Example: A Simplified Income Statement

Table A.3 A simplified income statement.

Income Statement (Year)	Amount ($)
Revenue	2,000,000
COGS	1,200,000
Gross profit	800,000
Operating expenses	400,000
Operating profit (EBIT)	400,000
Interest	50,000
Taxes	70,000
Net income	280,000

Interpretation: This company earned $280,000 in net profit on $2 million in revenue, resulting in a 14 percent net margin.

Why It Matters

The income statement reveals how efficiently a company turns revenue into profit. It highlights the impact of production costs, operating expenses, and financing decisions on overall performance. Managers use it to evaluate profitability, spot trends, and guide resource allocation. Investors and stakeholders use it to assess business viability and return potential.

Statement of Cash Flows

Profit is great, but cash is king. The cash flow statement shows why profit and cash are not the same. It tracks actual cash movement and divides activity into three categories:

- Operating activities: cash received from customers minus cash paid to suppliers, employees, and for other operating costs
- Investing activities: cash spent on long-term assets like equipment, buildings, or acquisitions or that is received from selling those assets
- Financing activities: cash received from loans or issuing shares and cash paid to repay debt or distribute dividends

Table A.4 provides an example of a simplified cash flow.

Worked Example: Simplified Cash Flow

Table A.4 Simplified cash flow.

Cash flow statement (Year)	Amount ($)
Net income	280,000
+ Depreciation	40,000
– Increase in inventory	(50,000)
– Increase in receivables	(30,000)
+ Increase in payables	20,000
Net cash from operations	260,000
– Capital expenditures	(100,000)
– Loan repayment	(80,000)
Net change in cash	80,000

Interpretation: Despite earning $280,000 in profit, the company's operating cash flow was slightly lower at $260,000 after adjusting for working capital changes. This difference was mild but illustrates that common effect expansion can quickly affect cash flow.

Why It Matters

The cash flow statement reveals whether the business can pay its bills, fund growth, or return value to shareholders. Substantial profits mean little if cash is tied up in receivables or excess inventory. Investors often focus on cash from operations as a signal of financial health. For managers, efficiency improvements usually show up in cash flow before they appear in reported profit.

How the Statements Fit Together

The three financial statements are tightly connected:

- The P&L statement reports profit, which is recorded in retained earnings under equity on the balance sheet.
- The cash flow statement starts with net income from the P&L and adjusts for noncash items and changes in working capital to show actual cash flow.
- Changes in balance sheet accounts, such as receivables, inventory, and payables, help explain why cash differs from reported profit.

Big Picture

- The balance sheet shows what the company owns and what it owes at a given point in time.
- The P&L shows revenue and expenses over a period.
- The cash flow statement indicates whether the company can pay its bills.

Worked Example: Following the Flow

A company records $500,000 in revenue and $350,000 in expenses, for a net income of $150,000.

On the balance sheet, cash does not increase by $150,000 because working capital uses of cash; $40,000 is tied up in receivables and $20,000 is tied up in new inventory.

The cash flow statement shows net cash from operations equal to $90,000.

Although simple, the example above illustrates why decision-makers consistently review all three statements together.

PROFITABILITY AND PERFORMANCE METRICS

Financial statements capture the details, but managers need concise indicators to assess whether the business is healthy, improving, or at risk. Profitability metrics and performance ratios simplify thousands of line items into a handful of meaningful numbers that support faster, smarter decisions.

Defining Noncash and Financing Terms

Please see these clarifications of the terms used here:

- Depreciation: allocation of the cost of physical assets (e.g., equipment) over their useful life

- Amortization: similar to depreciation but applied to intangible assets (e.g., software, patents)
- Taxes: refers to income taxes, which reduce net profit and depend on tax jurisdiction and strategy
- Interest: financing cost paid on debt

Profit

Profit is not one number. It comes in layers, each telling a different story about performance.

Gross Profit

Formula: Revenue – COGS

Focus: shows the revenue remaining after covering cost of goods sold.

Operating Profit (EBIT)

Formula: Gross profit – Operating expenses

Focus: reflects profitability after COGS and operating expenses, but before interest and taxes

Net Profit (Net Income)

Formula: Operating profit – (Interest + Taxes)

Focus: the "bottom line"—what remains for owners after all financing and tax obligations

Worked Example: Profit Levels

Suppose a company has:

Revenue: $2,000,000

COGS: $1,200,000

Operating expenses: $400,000

Interest: $50,000

Taxes: $70,000

Step by step:

Gross profit = $2,000,000 – $1,200,000 = $800,000

Operating profit = $800,000 – $400,000 = $400,000

Net profit = $400,000 – ($50,000 + $70,000) = $280,000

Margins

Margins express profit as a percentage of revenue, making it easier to compare performance across periods or companies of different sizes.

Gross Margin

Formula: (Gross profit ÷ Revenue) × 100

Focus: shows the revenue remaining after covering cost of goods sold

Operating Margin

Formula: (Operating profit ÷ Revenue) × 100

Focus: reflects profitability after overhead and SG&A, but before financing and taxes

Net Margin

Formula: (Net profit ÷ Revenue) × 100

Focus: indicates the percentage of revenue that becomes profit after all expenses

Worked Example

Using the example above:

Gross margin = 800,000 ÷ 2,000,000 = 40%

Operating margin = 400,000 ÷ 2,000,000 = 20%

Net margin = 280,000 ÷ 2,000,000 = 14%

Interpretation of ratios is always dependent on industry and structure but could suggest the following:

- 40 percent gross margin indicates strong production economics.
- 20 percent operating margin shows the company runs lean on overhead.
- 14 percent net margin is healthy after financing and taxes.

Margins are a robust measure because they normalize profit, allowing a $10 million company and a $100 million company to be compared directly.

Earnings Before Interest, Taxes, Depreciation, and Amortization (EBITDA)

EBITDA is a popular internal measure that strips out nonoperating and noncash expenses to show core operating performance.

Formula: EBITDA = Net income + Interest + Taxes + Depreciation + Amortization

Note: EBITDA is not reported on the income statement and is not defined under Generally Accepted Accounting Principles (GAAP) or International Financial Reporting Standards (IFRS). It should always be calculated and disclosed clearly.

Investors often use it to compare operating performance across firms in the same industry because these figures don't reflect daily operations but still affect reported profit. Stripping them out on an EBITDA basis gives a clearer view of performance before financial structure and accounting rules come into play.

Worked Example

Using the same income statement and adding

Depreciation = $40,000

Amortization = $0 (not specified)

Then

EBITDA = 280,000 + 50,000 + 70,000 + 40,000 = 440,000

Interpretation: EBITDA shows the business generates $440,000 before noncash charges and financing effects.

EBITDA = 440,000

Interpretation: While net income is $280,000, EBITDA suggests the core business throws off $440,000 before financing and noncash charges.

Contribution Margin

While gross and net margins show profitability at the company level, contribution margin looks at the per-unit economics of a product or service.

Formula: Sales price – Variable cost per unit

Contribution margin ratio = (Sales – Variable costs) ÷ Sales

Worked Example

Sales price per unit = $50

Variable cost per unit = $30

Contribution margin per unit = $20

Contribution margin ratio = 20 ÷ 50 = 40%

Interpretation: For every $1 of sales, $0.40 is available to cover fixed costs and profit.

Break-Even Analysis

Break-even analysis uses the contribution margin to find the sales volume required to cover all fixed costs.

Formula: Breakeven sales = Fixed costs ÷ Contribution margin ratio

Worked Example

Fixed costs = \$500,000

Contribution margin ratio = 40%

Break-even sales = \$500,000 ÷ 0.40 = \$1,250,000

Interpretation: The company must sell \$1,250,000 before it earns a single dollar of profit. Any revenue beyond this contributes directly to profit.

Return Metrics

These ratios measure how efficiently the company converts its resources into profit.

ROI = Net profit ÷ Investment cost

Above is a common form of ROI used for evaluating capital projects or new initiatives

Return on assets (ROA) = Net income ÷ Total assets

Measures how effectively assets are used to generate earnings

Return on equity (ROE) = Net income ÷ Shareholders' equity

Indicates the return owners earn on their capital

Worked Example

If the company has net income of \$280,000 and equity of \$750,000:

ROE = \$280,000 ÷ \$750,000 = 37% — a strong result

Economic Value Added (EVA)

Most metrics stop at profit. EVA goes further by asking whether we have earned enough to cover the cost of capital.

Formula: EVA = Net operating profit after taxes – (Capital × Cost of capital)

Interpretation:

If EVA > 0, the business is generating value above the cost of capital.

If EVA < 0, even if accounting profit is positive, the company is destroying shareholder value.

Worked Example

Net operating profit after taxes (NOPAT) = \$500,000

Capital invested = \$2,000,000

* Weighted average cost of capital (WACC) = 10%

EVA = \$500,000 – (\$2,000,000 × 0.10) = \$500,000 – \$200,000 = \$300,000

Interpretation: This shows the company created \$300,000 in actual economic value. EVA is increasingly popular among investors and boards because it ties performance to long-term enterprise value. EVA is not shown on financial statements. It must be calculated separately and is often used in board-level or investor analysis.

Example Takeaways

- Profit is multilayered. Gross, operating, and net profit each reflect different levels of cost.
- Margins convert profit to percentages, making performance easier to compare.
- EBITDA is a widely used proxy for operating performance but not part of standard financial reports.
- Contribution margin and break-even analysis support pricing and short-term decision-making.
- Return ratios (ROI, ROA, ROE) and EVA assess how well the business uses its capital.
- Some metrics are visible on financial statements; others must be calculated but are just as important for strategic decisions.

MANAGEMENT ACCOUNTING FOR DECISION-MAKING

Financial accounting looks backward. It records and reports historical data, allowing investors, regulators, and lenders to evaluate performance.

Management accounting looks forward. It organizes cost and performance data to help managers make better decisions about operations, projects, and resource allocation. The focus is not on compliance but on relevance: What information will help leaders take action?

Budgeting and Forecasting

Budgeting translates strategy into numbers. It sets expectations for revenue, costs, and profit and then serves as the benchmark for performance. Forecasting updates those numbers as conditions change.

- Static budgets: fixed at the start of the year
- Flexible budgets: adjust for actual activity levels
- Rolling forecasts: continuously updated, often projecting twelve months ahead

Budgets are not just controls. They are planning tools. For a quality leader, linking improvement initiatives to budget forecasts shows alignment with company goals.

Variance Analysis

Variance analysis compares performance with the budget or forecast. It shows how far performance falls short of planned performance.

Types of variances:

- Favorable (F): better than budget (higher revenue, lower cost)
- Unfavorable (U): worse than budget

Worked Example: Variance

Budget: $200,000 in monthly production costs

Actual: $210,000

Variance = $210,000 – $200,000 = $10,000 U (unfavorable).

Breakdown:

Higher raw material usage = $6,000 U

Higher overtime = $3,000 U

Higher energy = $1,000 U

Interpretation: Material usage drove most of the variance. Identifying material variance prompts managers to investigate scrap or process inefficiencies.

Overhead Allocation

Allocating overhead is one of management accounting's biggest challenges. Since indirect costs cannot be traced to a single product or customer, they must be allocated using a cost allocation base.

Standard allocation bases include

- direct labor hours,
- machine hours,
- units produced, and
- activity drivers (used in activity-based costing [ABC]).

Using hours—machine hours, labor hours, or units produced—overhead is distributed at a specific rate. Companies typically select the basis that is most applicable to their operations, though in some cases they may use two or even all three. The logic is straightforward: Costs are spread according to the chosen driver.

ABC works a little differently. Instead of relying on a single driver, it links costs to the specific activities that generate them.

ABC: Helpful Tool, Not a Requirement

ABC traces overhead more accurately by linking costs to the activities that cause them, such as the following:

- The number of machine setups may drive setup costs.
- The number of inspections may drive quality costs.

ABC avoids distorting product profitability when products differ in complexity or resource usage. However, it can be complex to maintain.

Working with Accounting, Not Around It

When an organization decides to track and analyze quality costs, it often turns to the accounting department with requests like, "Find all our quality-related costs" or "Break down our spending on investments for prevention, monitoring, waste, and downstream consequences." However, traditional accounting doesn't categorize expenses this way. Instead, costs are recorded in general accounts that don't directly reflect quality-related expenditures.

To bridge this gap, a quality specialist and an accountant must work together to identify and allocate quality costs within existing financial structures. Some cases are straightforward—for instance, an inspection department's budget typically falls under monitoring costs. However, many quality-related expenses, such as scrap and rework, are buried within standard cost categories and require a structured approach to identify them.

How ABC Improves Allocation Accuracy

ABC is widely recognized for improving cost allocation and financial decision-making. Yet, despite its benefits, adoption remains low. Researchers and managers agree that implementing ABC requires significant effort, which has slowed its widespread adoption for decades. However, for finance professionals and quality practitioners looking to refine their cost analysis, ABC remains a valuable tool.

Applying ABC to Reveal Hidden Costs

ABC enhances quality cost analysis by identifying hidden quality costs embedded in overhead expenses; accurately allocating costs to reflect actual resource consumption; highlighting cost-saving opportunities through improved visibility; distinguishing high- and low-performing areas within an organization; tracking quality cost trends to measure improvement efforts; and eliminating non-value-added activities to improve efficiency. By tracing cost drivers back to their root causes, companies can estimate the financial impact of quality improvements. For instance, if internal waste costs due to damage total $10,560, and corrective actions can reduce damage by 75 percent, the projected savings justify the investment in prevention.

ABC Adaptation and Value

While ABC adoption remains below 50 percent, it offers clear strategic benefits for businesses willing to invest in its implementation. For quality-focused professionals, understanding ABC principles provides a competitive edge in cost management and financial analysis. Traditional cost accounting can support quality cost analysis, but ABC offers a more precise and practical approach. By ensuring that overhead costs—often a significant portion of total expenses—are accurately assigned, ABC helps organizations make better financial decisions, improve quality management, and enhance operational efficiency.

Time-Driven Activity-Based Costing (TDABC)

While ABC provides more accurate overhead allocation, it is often criticized as complex and resource intensive. TDABC is a streamlined version of ABC. It replaces dozens of drivers with two simple inputs:

- Capacity cost rate: cost of supplying capacity per unit of time (e.g., $50 per machine hour)
- Time required: time to perform each activity (e.g., 0.5 hours per setup)

Formula: Cost of activity = (Capacity cost rate × Time required)

Advantages:

- Simpler to update and maintain
- Scales easily as activities change
- Allows managers to see unused capacity and cost implications

Example:

If a machine costs $100,000 per year to operate and is available for 2,000 hours, the capacity cost rate = $50/hour.

Setup activity = 0.5 hours → $25 per setup

Inspection activity = 0.2 hours → $10 per inspection

TDABC makes overhead allocation more transparent. Instead of spreading costs across broad bases, managers can see exactly what activities cost and whether resources are being fully used. The clarity of numbers in this example—$25 per setup, $10 per inspection—shows detail that would be buried or distorted in traditional systems. This shift turns overhead allocation from an accounting exercise into a practical tool for continuous improvement.

Decision Orientation

The hallmark of management accounting is actionability. Reports are not ends in themselves; they exist to guide choices:

- Should we continue making a component or buy it from a supplier?
- Should we keep all product lines or drop the low-margin ones?
- Should we allocate scarce capital to Project A or Project B?

In each case, management accounting provides the relevant costs, forecasts, and performance data to evaluate alternatives.

Key Takeaways

- Classifying costs correctly determines which levers managers can pull.
- Budgets and forecasts provide the baseline for evaluating results.
- Variance analysis highlights the fundamental drivers of performance gaps.
- Overhead allocation, primarily through ABC or TDABC, reveals the profitability of products or services.
- Cost centers and profit centers represent different accountability models.
- Above all, management accounting is about decisions, not reporting.

PROJECT EVALUATION PRINCIPLES

Management accounting equips managers with the numbers to run the business day-to-day. But when deciding whether to approve a new initiative, whether a machine investment, a supplier-quality program, or a training rollout, managers need methods that weigh costs against future benefits.

Three tools dominate in practice: NPV, internal rate of return (IRR), and payback period. Before looking at them, we need the foundation: the time value of money (TVM).

TVM

At the heart of project evaluation lies one principle: A dollar today is worth more than a dollar tomorrow. Why?

- A dollar today can be invested to earn a return.
- Inflation erodes the purchasing power of future dollars.
- Future outcomes are uncertain, so value today is preferred.

Equations: where PV = present value, FV = future value, r = interest rate (per period), and n = number of periods.

FV

FV = PV × (1 + r)^n

PV = FV ÷ (1 + r)^n

Worked Example: TVM

If you expect to receive $10,000 five years from now, and the discount rate is 6 percent

PV = $10,000 ÷ (1.06)^5 = $10,000 ÷ 1.3382 = $7,473

Interpretation: That future $10,000 is worth only $7,473 today.

NPV

NPV applies a discount rate to project cash flows. It asks: After bringing all future inflows back to today and subtracting the initial cost, is the project worth more than zero?

Formula: NPV = Σ [Cash flow_t ÷ (1 + r)^t] – Initial investment

Rule: If NPV > 0, accept. If NPV < 0, reject.

Focus: measures the dollar value a project adds (or destroys) after discounting future cash flows to today.

Worked Example: NPV

The project requires $100,000 up front. It saves $30,000 annually for five years.

Discount rate = 5%.

Step 1: Calculate PV of annual savings

Year 1 PV = 30,000 ÷ 1.05 = 28,571

Year 2 PV = 30,000 ÷ 1.05^2 = 27,211

Year 3 PV = 30,000 ÷ 1.05^3 = 25,915

Year 4 PV = 30,000 ÷ 1.05^4 = 24,681

Year 5 PV = 30,000 ÷ 1.05^5 = 23,506

Total PV inflows = 129,884

Step 2: Subtract initial investment

NPV = 129,884 – 100,000 = 29,884

Interpretation: The project adds nearly $30,000 of value. It should be accepted.

Strength: works across projects of different sizes and timing; preferred by finance.

Limitation: requires selecting an appropriate discount rate.

IRR

Definition: The discount rate that makes NPV = 0.

Rule:

- If IRR > company's required return (hurdle rate), accept the project.
- If IRR < hurdle rate, reject it.

Focus: expresses a project's return as a percentage, useful for ranking investments.

IRR helps rank projects because it expresses results as a percentage. But it can be misleading when projects differ in scale or timing or when they have "nonnormal" cash flows (e.g., alternating inflows and outflows).

Worked Example: IRR

For the project above ($100,000 upfront, $30,000 annual savings, 5 years):

IRR = 15.2%

Interpretation: If the company's required return is 10 percent, this project easily clears the hurdle.

Limitations: IRR is popular but not foolproof. It can mislead when

- projects differ in scale (a smaller project may show a higher IRR but deliver far less total profit),
- cash flow timing differs (early returns inflate IRR even if the total value is smaller),
- cash flows are nonnormal (multiple outflows and inflows can create multiple IRRs or none at all), and
- the reinvestment assumption is unrealistic (it assumes that interim cash flows are reinvested at the IRR).

Modified IRR (MIRR)

MIRR addresses the reinvestment flaw in IRR. IRR has flaws when cash flows alternate between inflows and outflows, so MIRR fixes this by making more realistic assumptions:

- Positive cash flows are reinvested at the firm's cost of capital.
- Negative cash flows are financed at the firm's borrowing cost.

MIRR produces a single, realistic return figure.

Equation: MIRR = (FV of inflows reinvested at reinvestment rate ÷ PV of outflows at finance rate)^(1/n) – 1

Strength: provides a single, realistic return figure.

Limitation: less commonly used in everyday practice than IRR.

Takeaway: A project with uneven cash flows may yield two IRRs (8 and 22 percent), complicating interpretation. MIRR instead produces a single figure (e.g., 12 percent), which can be compared directly with the company's hurdle rate.

Payback Period

Formula: Payback = Initial investment ÷ Annual cash inflow

Focus: time required to recover the original investment.

Strength: simple and highlights liquidity risk.

Limitation: ignores TVM and benefits after payback.

Worked Example: Payback

Same project: $100,000 investment, $30,000 annual savings

Payback = 100,000 ÷ 30,000 = 3.33 years

Interpretation: The project breaks even after just three years; any benefits beyond that are profit.

Comparing the Three Tools

- NPV: best measure of value creation; works across different project sizes and cash flow patterns
- IRR: intuitive percentage measure but can mislead when projects differ in scale, timing, or have irregular cash flows
- MIRR: improves on IRR by correcting reinvestment assumptions and eliminating multiple results
- Payback: quick check on liquidity and risk but incomplete on its own

Best practice: Always calculate NPV. Use IRR or MIRR for quick comparisons, and payback when short-term cash needs matter.

Worked Example: Side-by-Side Comparison

Project Alpha: Invest $150,000 upfront. Expect $50,000 in annual savings for 4 years. Discount rate = 8%.

NPV: PV of inflows ≈ 165,000. Subtract 150,000. NPV = +15,000
IRR: ≈ 11%
Payback: 150,000 ÷ 50,000 = 3 years

Interpretation: NPV shows it adds $15,000 of value. IRR clears the 8 percent hurdle rate. Payback is reasonable at 3 years.

Interpretation: As expected with a simple project, all three measures align. Definitely, this is a good project for an organization with an 8 percent discount rate. If the discount rate or required rate of return were 22 percent, the conclusion would be different.

Practical Tips

- *Always calculate NPV—it is the most reliable indicator of value.*
- *Use IRR to communicate with nonfinance managers who think in percentages.*

- *Use payback when liquidity is tight or management wants reassurance that funds will be recovered quickly.*
- *Run sensitivity tests (covered in Sensitivity Analysis subsection of the Risk and Uncertainty section) to see how results change if savings are smaller, costs are higher, or discount rates shift.*

RISK, TREASURY, AND CAPITAL RATIONING

On paper, project evaluation looks simple. If NPV is positive, fund the project. If IRR exceeds the hurdle rate, accept it. If payback is short, feel reassured.

In practice, things are less tidy. Forecasts are uncertain, cash flow timing matters, and resources are limited. Managers must weigh financial metrics against practical constraints.

Sunk Costs and Decision Biases

A sunk cost is money already spent that cannot be recovered. Rationally, it should not affect future decisions. Only incremental costs and benefits matter.

Yet many managers fall into the sunk cost fallacy: continuing a project because "we've already spent too much to stop," even when forecasts no longer support it.

Here is some practical advice:

- Focus only on future costs and benefits.
- State explicitly in business cases: "Past spending is not relevant to this decision."

Opportunity Cost

Every dollar spent on one project is a dollar unavailable for another. Even if a project has a positive NPV, it may not be the best choice if another use of funds offers more value. Managers must compare across functions and time horizons.

Treasury Management and Short-Term Liquidity Risk

Profit is not the same as cash. A business can be profitable on paper but still face a cash crunch if inflows and outflows do not align. Common issues include the following:

- Customers delay payment, inflating receivables.
- Inventory builds faster than sales, tying up cash.
- Capital purchases drain reserves.

When business conditions are stable, treasury management often happens quietly in the background. It is always present, but during a crisis or a period of rapid expansion, treasury managers suddenly become highly visible. The earlier example of a company showing respectable profits but consuming cash excessively illustrates the point. Profitability on paper is not enough if the business cannot meet its obligations in real time. Risk of default on obligations is why treasury management—the control of cash inflows and outflows—matters just as much as profit.

Table A.5 gives an example of a company that is cash poor yet profitable.

Worked Example: Profitable but Cash Poor

Table A.5 Profitable but cash poor.

Item	Amount ($)
Net income	500,000
+ Increase receivables	(300,000)
+ Increase inventory	(250,000)
Net cash from operations	(50,000)

Despite a robust profit, the company consumed cash and may struggle to pay suppliers.

Implication: A project with a short payback or immediate cash benefits may be prioritized over one with a higher NPV but delayed benefits.

Risk and Uncertainty

Quality folks are familiar with probability and uncertainty. Likewise, forecasts rarely play out exactly as expected. Managers use two tools to test resilience: sensitivity analysis and scenario analysis. Table A.6 gives an example of data for a sensitivity analysis.

Sensitivity Analysis

Change one variable at a time.

Example:

If annual savings fall from $100,000 → $80,000 → $60,000,
NPV drops from +$50,000 → +$10,000 → –$30,000.

Table A.6 Sensitivity analysis.

Annual savings ($)	NPV ($)
100,000	+50,000
80,000	+10,000
60,000	–30,000

Scenario Analysis

Combine assumptions into full cases.
Example:

Best case = higher sales, lower costs
Base case = original forecast
Worst case = delays and higher costs

Interpretation: A project may look acceptable under base assumptions but fragile under downside scenarios.

Capital Rationing

Textbook theory says that all projects with a positive NPV should be accepted. Mathematically and logically, this is correct because each increases shareholder value (even when accounting for the costs of securing internal or external capital). The proper approach, technically and academically, is to approve every project that meets the hurdle rate and appropriately accounts for risk.

In practice, this is seldom the case. Companies face limited budgets, constrained management capacity, and concerns about risk concentration. Choices must be made, even among positive NPV projects. (See Table A.7.)

Projects may be rejected for such reasons as follow, despite positive NPV:

- Capital budgets are capped (e.g., $10 million available vs. $25 million in proposals).
- Liquidity pressure favors faster payback.
- Priorities compete across functions (operations vs. marketing vs. IT).
- Risk is concentrated (too many projects reliant on a single supplier, market, or assumption).

Worked Example: Capital Rationing

The company has $500,000 available for investment and three proposed projects.

Table A.7 Capital rationing.

Project	Investment ($)	NPV ($)	Payback (yrs)
A	200,000	40,000	3
B	300,000	60,000	4
C	250,000	35,000	2

All are positive NPV. But only A + C can be funded within budget ($450,000), with a combined NPV of +$75,000, which is higher than choosing B alone (+$60,000).

Implication: Ranking and selection become necessary. In the real world, firms rarely fund every positive-NPV project—they fund the mix that best balances value, risk, and resource constraints.

Practical Takeaways

- Ignore sunk costs when making decisions.
- Always consider opportunity cost.
- Profit does not equal cash; liquidity discipline matters.
- Use sensitivity and scenario analysis to stress-test assumptions.
- Not every positive NPV project gets funded. Prioritization is key.

ACCOUNTING STANDARDS AND PRINCIPLES

Financial statements are not prepared on an ad hoc basis. They follow established rules so that investors, lenders, regulators, and managers can rely on the numbers and compare across companies. These rules are known as *accounting standards.*

Standards Bodies

Two major systems dominate globally:

- GAAP: Used primarily in the United States and established by the Financial Accounting Standards Board (FASB), GAAP is often described as a rules-based system, with detailed guidance for specific situations.
- IFRS: Used in most other countries and set by the International Accounting Standards Board (IASB), IFRS is considered principles based, emphasizing broader concepts that require interpretation in practice.

Although GAAP and IFRS differ in style, both aim to produce financial reports that are relevant, reliable, and comparable.

Core Principles

The exact number of principles depends on how they are grouped and which system you reference. GAAP is traditionally taught in ten sections, while IFRS highlights fewer, broader concepts. For most purposes, six core ideas capture the common ground:

- Accrual principle: Record revenues when earned and expenses when incurred, not when cash moves.
- Matching principle: Align expenses with the revenues they help generate. For example, warranty costs are recorded in the same period as the sale, not when claims are paid.

- Consistency principle: Once a method is chosen (e.g., depreciation), apply it consistently unless a justified change improves accuracy.
- Prudence/conservatism: Avoid overstating assets or income; recognize expenses and liabilities promptly.
- Going concern assumption: Assume the business will continue operating, unless evidence suggests otherwise.
- Materiality: Focus on information that would influence a decision; immaterial items may be grouped or omitted.

Why Results May Differ

Even with shared principles, GAAP and IFRS sometimes produce different outcomes because of their underlying philosophies:

- Flexibility: IFRS allows more managerial judgment in applying principles, while GAAP prescribes more specific methods.
- Asset valuation: IFRS permits the revaluation of certain assets to fair market value; GAAP typically requires assets to be carried at historical cost.
- Inventory methods: GAAP allows such methods as LIFO (last in, first out). IFRS prohibits LIFO, which can create differences in reported costs and margins during inflationary periods.
- Impairments: IFRS allows reversals of impairment losses if conditions improve; GAAP does not.

For managers outside finance and accounting, the point is not to memorize every rule but to recognize that the principles provide consistency within each system and that both frameworks point in the same overall direction.

Implications for Managers

For quality professionals and other managers, the key insight is that financial reporting often smooths or delays recognition of operational impacts:

- A cost-saving project may reduce cash outflow immediately, but accrual and matching rules can spread the benefit across several periods.
- A warranty provision may reduce profit in the year of sale, even if cash claims occur years later.
- Depreciation allocates the cost of equipment over its useful life, even though the cash was spent up front.

The matching principle, in particular, can feel counterintuitive or even deceptive, but like all the principles, it is intentional. The goal of standards is to present a fair and consistent picture of performance over time, not just a snapshot of cash movement.

Why Accounting Matters

When you make the case for quality initiatives, understanding these principles helps you:

- Anticipate when results will appear in financial statements.
- Explain why P&L benefits may not align perfectly with cash savings.
- Communicate effectively with finance colleagues about costs, provisions, and capital spending.

Key Takeaway

Whether under GAAP or IFRS, accounting standards are not designed to frustrate managers. They ensure consistency and comparability. For quality professionals, the practical value lies in knowing enough to align improvement projects with how results will be reported and perceived by executives and shareholders.

VALUATION: LINKING PROJECTS TO ENTERPRISE VALUE

So far, we've covered costs, financial statements, profitability, management accounting, project evaluation, and risk. These tools help managers decide whether an initiative is worthwhile.

Executives, boards, and investors often step back and ask a higher-level question: How does this affect the business's value?

That question leads us to valuation.

What Drives Company Value?

At its core, a company's value depends on its ability to generate future cash flows for its owners and lenders.

- A company with strong, reliable, and growing cash flows is worth more than one with weak or uncertain flows.
- Operational improvements, including COQ initiatives, matter because they can
  - Increase revenues: External failure costs (e.g., defects reaching customers) hurt satisfaction and loyalty. Preventing them supports repeat sales and brand value.
  - Reduce costs: Internal failure (scrap, rework) and appraisal costs drain resources. Reducing them improves margins and frees capacity.
  - Free up working capital: Quality-driven process improvements (e.g., shorter cycle times, fewer inspection delays) reduce inventory and receivables.
  - Reduce risk: Prevention efforts lower compliance failures, warranty claims, and the risk of regulatory fines—all of which investors heavily discount when valuing a business.

In other words, COQ initiatives are not just "quality projects." They are value drivers because they directly impact revenue growth, cost structure, capital efficiency, and risk—the four levers investors and executives use to judge enterprise value.

Three Valuation Approaches

Finance professionals use three main approaches to estimate value:

Discounted Cash Flow (DCF)

- An intrinsic valuation method, DCF estimates the present value of future free cash flows
- DCF is best suited for linking operational improvements directly to enterprise value, since it highlights how revenue, cost, and risk drivers translate into long-term cash generation

Relative Valuation (Multiples and Comparables)

- Relative valuation is based on how similar companies are valued in the market.
- Multiples apply industry averages (e.g., 8 × EBITDA, 2 × revenue).
- Comparables are based on transactions or peer benchmarks.
- Fast and widely used by investors, bankers, and analysts, relative valuation is limited because it assumes "average" market ratios apply equally to all businesses.

Contingent Claim Valuation (CCV)

- Also called "real options," this method values flexibility—the ability to delay, expand, or abandon projects as conditions change.
- More advanced and data heavy, CCV is practical when uncertainty is high or outcomes depend on external shifts, such as regulation, demand volatility, or technology adoption.

Multiples and comparables are widely used, but they are blunt instruments. They support quick external comparisons but lack the insight managers need to connect specific improvements to value. CCV is more nuanced but general for exceptional cases, including high uncertainty. DCF, while more intrinsic, shows how pulling operational levers, raising revenue, cutting costs, or reducing risk translates into enterprise value.

Free Cash Flow (FCF)

The cornerstone of valuation is FCF: the cash a company generates after covering its operating expenses and necessary capital expenditures. It represents the money available to repay debt, return to shareholders, or reinvest in growth.

Formula (simplified): FCF = Operating cash flow – Capital expenditures

Discounted Cash Flow (DCF)

The power of DCF lies in its ability to make future performance concrete. Imagine you are standing five or ten years ahead, looking back:

1. Start with forecasting. What level of free cash flow will the business generate each year? Include the effects of revenues, costs, working capital, and capital expenditures. For quality leaders, consider how reducing scrap, shortening lead times, or improving customer retention directly affects those numbers.
2. Adjust for risk and time. A dollar today is more valuable than a dollar tomorrow, partly because of inflation but also because of uncertainty. To bring future dollars back to present value, we discount them using the company's weighted average cost of capital (WACC). WACC blends the cost of debt (what banks charge) and equity (the return shareholders expect).
3. Account for the long term. Businesses do not just stop after the forecast horizon. A terminal value captures all future years beyond, typically assuming steady growth.
4. Add it all together. The present value of forecast cash flows plus the PV of the terminal value equals the company's enterprise value.

In other words, DCF turns operating decisions, pricing, quality, working capital, and risk control into a single number representing the company's current value.

Equation (simplified):

$$\text{Enterprise Value} = \Sigma\,[FCF\_t \div (1 + r)^\wedge t] + [\text{Terminal value} \div (1 + r)^\wedge n]$$

Where:

t = year in the forecast

r = discount rate (WACC)

Terminal value often = FCF_(n+1) ÷ (r − g), where g = assumed long-term growth rate

Worked Example: A Simple DCF

Suppose a company expects free cash flow of $200,000 per year for the next five years. After that, it's assumed to grow at 2 percent per year. Discount rate (WACC) = 8%.

Step 1: Forecast cash flows.

Year 1–5: 200,000 each.

Step 2: Discount them.

Year 1 PV = 200,000 ÷ 1.08 = 185,185
Year 2 PV = 200,000 ÷ (1.08)^2 = 171,468
Year 3 PV = 158,768
Year 4 PV = 147,014
Year 5 PV = 136,122

Total PV (Years 1–5) = 798,557

Step 3: Terminal value.

Year 6 cash flow = 200,000 × (1 + 0.02) = 204,000
Terminal value = 204,000 ÷ (0.08 – 0.02) = 3,400,000
Discounted back to present: 3,400,000 ÷ (1.08^5) = 2,312,446

Step 4: Enterprise value.

798,557 + 2,312,446 = 3,111,003

Interpretation: Based on this forecast, the company is valued at approximately $3.1 million.

Linking Operational Improvements to Valuation

Now imagine a quality initiative saves $200,000 in costs every year, permanently. How does that affect value?

Annual FCF increases by $200,000.

Using the same 8 percent discount rate and 2 percent growth assumption:

Terminal value impact = 200,000 ÷ (0.08 – 0.02) = 3,333,333

Present value ≈ 2.27 million.

So one sustained operational improvement could raise enterprise value by more than $2 million.

CCV

Not all opportunities can be captured with static DCF. Some projects resemble options—rights but not obligations to act later—such as the following:

- Delay investment until uncertainty clears (option to wait).
- Expand if demand grows (option to expand).
- Abandon if results disappoint (option to exit).

Traditional NPV ignores these flexibilities. CCV (or real options valuation) borrows from financial option pricing to value them.

Brief example: A project with a slightly negative NPV may become attractive if management holds an option to expand in year 2 if demand is strong. CCV adds that option value, turning an apparent no into a conditional yes.

While CCV is more advanced, the key lesson is that flexibility itself has value—a point often overlooked in traditional analysis.

Multiples and Comparables

Multiples and comparables estimate a firm's value by comparing a company to others in the same industry. It usually involves applying multiples (such as EV/EBITDA or price-to-sales) to financial results or using comparables like growth rates and per-customer metrics.

This approach is popular because it is quick, market based, and grounded in observable data. Investors and analysts often rely on it for benchmarking.

Examples across industries include the following:

- Manufacturing: EV/EBITDA multiples are common since they approximate operating cash flow.
- Retail: Same-store sales growth and price-to-sales ratios are standard comparables.
- Software: Revenue per subscriber or monthly active users often drive valuation multiples.
- Education/training: Enrollment growth and per-student tuition revenue are standard benchmarks.

Relative valuation works well when improvements affect headline metrics that markets track—such as higher revenue, stronger customer retention, or margin expansion. But many quality initiatives focus on cost efficiency, which can be harder to see in multiples unless the savings are quite considerable and sustained. For project-level evaluation, DCF provides a clearer link.

Valuation Key Takeaways

- The value of a company rests on its ability to generate future cash flows.
- FCF is the central concept in valuation.
- DCF is the most direct and nuanced way to link operational improvements to enterprise value.
- Multiples and comparables provide useful shortcuts but should not substitute for deeper analysis.
- Flexibility, captured through CCV, can be valuable in its own right.
- A single, sustained improvement in cost, revenue, or working capital can multiply enterprise value.

- For quality professionals, *COQ initiatives directly align with these value levers*, strengthening the link between process improvement and shareholder value.

SUMMARY AND KEY TAKEAWAYS

This appendix has covered a wide arc, from the mechanics of cost structures to the fundamentals of valuation. While not every quality professional needs to be fluent in accounting or finance, understanding the language of money will make you far more effective in conversations with executives, finance teams, and even external stakeholders.

Stepping back, the essentials line up as follows:

- *Cost structures.* Manufacturing direct and indirect costs flow into COGS, while selling and general and administrative costs are reported separately.
- *Cost vs. profit centers.* Departments are measured differently depending on whether they generate revenue or only incur costs. Understanding this distinction helps frame performance discussions appropriately.
- *Financial statements.* The balance sheet shows what the company owns and owes. The P&L statement shows how much it earned over a period. The cash flow statement shows whether those earnings convert into liquidity. Together, they provide a comprehensive view of financial health.
- *Profitability metrics.* Margins, EBITDA, contribution margin, and break even reduce thousands of transactions into a few measures that reveal efficiency and performance.
- *Management accounting.* Budgets, forecasts, cost classifications, and variance analysis are tools for making better decisions, not just compliance exercises.
- *Project evaluation.* NPV, IRR, payback, and related methods weigh costs against benefits. Among them, NPV most directly captures value creation.
- *Risk and treasury.* Profit is not the same as cash. Sensitivity testing, liquidity management, and capital rationing remind us that survival and growth must be balanced.
- *Accounting standards.* GAAP and IFRS smooth timing differences and ensure consistency. Managers explain why reported results may not precisely align with operational improvements.
- *Valuation.* Ultimately, worth comes from the ability to generate free cash flow. DCF links quality projects directly to enterprise value, while multiples and comparables provide external benchmarks grounded in the same logic.

Why This Matters for Quality Professionals

For quality leaders, the message is simple: Finance is not a separate world. The work of reducing scrap, preventing failures, or improving cycle times directly affects the levers that determine enterprise value. By understanding financial concepts, you can

- translate COQ improvements into financial terms that resonate with executives,
- anticipate how initiatives will appear in financial statements,
- frame projects with NPV, IRR, or payback to show not just efficiency but financial return, and
- link operational excellence to shareholder value, positioning quality as a driver of business success.

To remain relevant now and in the future, quality must drive improvements that connect to leadership and the bottom line. No longer is it sufficient to just fix defects or reduce rework for quality's sake. It is about strengthening the business's financial engine—boosting cash flow, reducing risk, and creating lasting value.

Financial Fluency

As you advance in your career, the ability to "speak finance" will set you apart. It builds credibility, opens doors to strategic influence, and allows you to position quality as more than a technical discipline. Many finance professionals spend entire careers driving value without ever touching the product or process. Quality professionals occupy a rarer, more powerful position at the intersection of product, process, and value creation.

When you add financial fluency to that vantage point, you move from being a technical expert to being a business leader. The number in the box may be expected, but the understanding behind it can shape strategy, guide investment, and ensure that improvement creates value where it counts.

Appendix B

Detailed Description of Quality Cost Elements

Appendix B reflects the accumulated expertise of former Cost of Quality Committee members across many editions of this book. It is intended as a starter list of COQ sources commonly found in organizations and as a practical tool for beginning (or refining) a COQ study. For this edition, the Prevention-Appraisal-Failure (PAF) terminology used in prior editions has been revised to the current Investment-Monitoring-Waste-Downstream Consequence framework, while preserving the original intent and usability of the content.

This appendix identifies detailed quality cost elements in numerical sequence (see Figure B.1 for a summary) to support consistent classification and practical use. Not all elements apply to every organization; readers are expected to determine applicability based on their unique circumstances. The list isn't intended to be exhaustive, but to illustrate the types of costs typically included within each category and to help in assigning individual costs during actual application. If a significant cost aligns with any part of an element's general description, then it should be included. Because many quality-related activities involve personnel from multiple functions, no attempt is made to assign costs to specific departments, recognizing that organizational structures vary.

1.0	**INVESTMENTS**		
1.1	Marketing/Customer/User	1.4	Operations (Manufacturing or Service) Quality Investments
1.1.1	Marketing Research	1.4.1	Operations Process Validation
1.1.2	Customer/User Perception Surveys/ Clinics	1.4.2	Operations Quality Planning
1.1.3	Contract/Document Review	1.4.2.1	Design and Development of Quality Measurement Control Equipment
1.2	Product/Service/Design Development		
1.2.1	Design Quality Progress Reviews	1.4.3	Operations Support Quality Planning
1.2.2	Design Support Activities	1.4.4	Operator Quality Education
1.2.3	Product Design Qualification Test	1.4.5	Operator SPC/Process Control

Figure B.1 Detailed quality cost element summary.

1.2.4	Service Design-Qualification	1.5	Quality Administration
1.2.5	Field Trials	1.5.1	Administrative Salaries
1.3	Purchasing Investments	1.5.2	Administrative Expenses
1.3.1	Supplier Reviews	1.5.3	Quality Program Planning
1.3.2	Supplier Rating	1.5.4	Quality Performance Reporting
1.3.3	Purchase Order Tech Data Reviews	1.5.5	Quality Education
1.3.4	Supplier Quality Planning	1.5.6	Quality Improvement
		1.5.7	Quality System Audits
		1.6	Other Quality Investments
2.0	**MONITORING COSTS**		
2.1	Purchasing Monitoring Costs	2.2.4	Process Control Measurements
2.1.1	Receiving or Incoming Inspections	2.2.5	Laboratory Support
	and Tests	2.2.6	Measurement (Inspection and Test)
2.1.2	Measurement Equipment		Equipment
2.1.3	Qualification of Supplier Product	2.2.6.1	Depreciation Allowances
2.1.4	Source Inspection and Control	2.2.6.2	Measurement Equipment Expenses
	Programs	2.2.6.3	Maintenance and Calibration Labor
2.2	Operations (Manufacturing or Service)	2.2.7	Outside Endorsements and
	Monitoring Costs		Certifications
2.2.1	Planned Operations Inspections, Tests,	2.3	External Monitoring Costs
	Audits	2.3.1	Field Performance Evaluation
2.2.1.1	Checking Labor	2.3.2	Special Product Evaluations
2.2.1.2	Product or Service Quality Audits	2.3.3	Evaluation of Field Stock and Spare Parts
2.2.1.3	Inspection and Test Materials		
2.2.2	Set-Up Inspections and Tests	2.4	Review of Test and Inspection Data
2.2.3	Special Tests (Manufacturing)	2.5	Miscellaneous Quality Evaluations
3.0	**WASTE COSTS**		
3.1	Product/Service Design Wastes (Internal)	3.3.1.1	Disposition Costs
3.1.1	Design Corrective Action	3.3.1.2	Troubleshooting or Failure Analysis Costs (Operations)
3.1.2	Rework Due to Design Changes	3.3.1.3	Investigation Support Costs
3.1.3	Scrap Due to Design Changes		
3.1.4	Production Liaison Costs	3.3.1.4	Operations Corrective Action
3.2	Purchasing Failure Costs	3.3.1.2	Operations Rework and Repair Costs

Figure B.1 Detailed quality cost element summary *(continued)*.

3.2.1	Purchased Material Reject Disposition Costs	3.3.2.1	Rework
3.2.2	Purchased material Replacement	3.3.2.2	Repair
	Costs	3.3.2	Operations Rework and Repair Costs
3.2.3	Supplier Corrective Action	3.3.3	Re-inspection/Re-test Costs
3.2.4	Rework of Supplier Rejects	3.3.4	Extra Operations
3.2.5	Uncontrolled Material Losses	3.3.5	Scrap Costs (Operations)
3.3	Operations (Product or Service) Waste Costs	3.3.6	Downgraded End-Product or Service
	Material Review and Corrective Action	3.3.7	Internal Waste Labor Losses
3.3.1	Costs	3.4	Other Waste Costs
		3.5	Office Wastes
		3.5.1	Meetings with No Good Outcomes
		3.5.2	Distractions (Good Minus Bad Distractions)
		3.5.3	Repetitive Email and Other
		3.5.4	Messaging
			Information Loss from Storage and
		3.5.5	Retrieval Issues
		3.5.6	Time Lost Due to Low Motivation
		3.5.7	Poor Results and/or Workflow Arrangements (Physical and Virtual)
4.0	**DOWNSTREAM CONSEQUENCE COSTS**		
4.1	Complaint Investigations/Customer or User Service	4.5	Liability Costs
		4.6	Penalties
4.2	Returned Goods	4.7	Customer/User Goodwill
4.3	Retrofit Costs	4.8	Lost Sales
4.3.1	Recall Costs	4.9	Other Downstream Consequence Costs
4.4	Warranty Claims		

Figure B.1 Detailed quality cost element summary (*continued*).

1.0 INVESTMENTS

Expenditures on all activities specifically designed to prevent poor quality in products or services.

1.1 Marketing/Customer/User

Expenditures incurred in the accumulation and continued evaluation of customer and user quality needs and perceptions (including feedback on reliability and performance) affecting their satisfaction with the company's product or service.

1.1.1 Marketing Research

The expenditures of the portions of marketing research devoted to the determination of customer and user quality needs—attributes of the product or service that provide a high degree of satisfaction.

1.1.2 Customer/User Perception Surveys/Clinics

Expenditures for programs designed to communicate with customers/users for the express purpose of determining their perception of product or service quality as delivered and used, from the viewpoint of their expectations and needs relative to competitive offerings.

1.1.3 Contract/Document Review

Expenditures incurred in the review and evaluation of customer contracts or other documents affecting actual product or service requirements (such as applicable industry standards, government regulations, or customer internal specifications) to determine the company's capability to meet the stated requirements, prior to acceptance of the customer's terms.

1.2 Product/Service/Design Development

Expenditures incurred to translate customer and user needs into reliable quality standards and requirements and manage the quality of new product or service developments prior to the release of authorized documentation for initial production. These costs are normally planned and budgeted and are applied to major design changes as well.

1.2.1 Design Quality Progress Reviews

The total expenditures, including planning, of interim and final design progress reviews, conducted to maximize conformance of product or service design to customer or user needs with regard to function, configuration, reliability, safety, producibility, unit cost, and as applicable, serviceability, inter-changeability, and maintainability. These formal reviews will occur prior to release of design documents for fabrication of prototype units or start of trial production.

1.2.2 Design Support Activities

The total expenditures for all activities specifically required to provide tangible quality support inputs to the product or service development effort. As applicable, design support activities include design document checking to assure conformance to internal design standards; selection and design qualification of components and/or materials required as an integral part of the end-product or service; risk analyses for the safe use of end-product or service; producibility studies to assure economic production capability; maintainability or serviceability analyses; reliability assurance activities such as failure mode and effects analysis and reliability apportionment; analysis of customer misuse and abuse potential; and preparation of an overall quality management plan.

1.2.3 Product Design Qualification Test

Expenditures incurred in the planning and conduct of the qualification testing of new products and major changes to existing products. Includes expenditures for the inspection and test of a sufficient quantity of qualification units under ambient conditions and the extremes of environmental parameters (worst case conditions). Qualification inspections and tests are conducted to verify that all product design requirements have been met or, when failures occur, to clearly identify where redesign efforts are required. Qualification testing is performed on prototype units, pilot runs, or a sample of the initial production run of new products. (Some sources consider this a monitoring cost.)

1.2.4 Service Design—Qualification

Expenditures incurred in the qualification or overall process proving of new service offerings and major changes to existing offerings. Involves planning for and performing a pilot or trial run using prototype or first production supplies as required. Includes detailed measurements or observations of each aspect of the service offering under normal and worst-case conditions, for a sufficient quantity of units or time as applicable, to verify consistent conformance to requirements, or to identify where redesign efforts are required. (Some sources consider this a monitoring cost.)

1.2.5 Field Trials

Expenditures for planned observations and evaluation of end-product performance in trial situations—usually done with the cooperation of loyal customers but also includes sales into test markets. At this stage of product or service life, a company needs to know much more than "Did it work?" or "Did it sell?" (Some sources consider this a monitoring cost.)

1.3 Purchasing Investments

Expenditures incurred to ensure conformance to requirements of supplier parts, materials, or processes and to minimize the impact of supplier nonconformances on the quality of delivered products or services. Involves activities prior to and after finalization of purchase order commitments.

1.3.1 Supplier Reviews

Expenditures for surveys to review and evaluate individual supplier's capabilities to meet company quality requirements. Usually conducted by a team of qualified company representatives from affected departments. Can be conducted periodically for long-term associations.

1.3.2 Supplier Rating

Expenditures for developing and maintaining, as applicable, a system to ascertain each supplier's continued acceptability for future business. This rating system is based on actual supplier performance to established requirements, periodically analyzed, and given a quantitative or qualitative rating.

1.3.3 Purchase Order Tech Data Reviews

Expenditures for reviews of purchase order technical data (usually by other than purchasing personnel) to assure its ability to clearly and completely communicate accurate technical and quality requirements to suppliers.

1.3.4 Supplier Quality Planning

Expenditures for planning for the incoming and source inspections and tests necessary to determine acceptance of supplier products. Includes the preparation of necessary documents and development expenses for newly required inspection and test equipment.

1.4 Operations (Manufacturing or Service) Quality Investments

Expenditures incurred in assuring the capability and readiness of operations to meet quality standards and requirements; quality control planning for all production activities; and the quality education of operating personnel.

1.4.1 Operations Process Validation

Expenditures on activities established for the purpose of assuring the capability of new production methods, processes, equipment, machinery, and tools to initially and consistently perform within required limits.

1.4.2 Operations Quality Planning

Expenditures for development of necessary product or service inspection, test, and audit procedures; monitoring documentation systems; and workmanship or appearance standards to assure the continued achievement of acceptable quality results. Also includes total design and development expenditures for new or special measurement and control techniques, gauges, and equipment.

1.4.2.1 Design and Development of Quality Measurement and Control Equipment

Expenditures for test equipment, engineers, planners, and designers; gauge engineers; and inspection equipment engineers, planners, and designers.

1.4.3 Operations Support Quality Planning

The total expenditures on quality control planning for all activities required to provide tangible quality support to the production process. As applicable, these production support activities include, but are not limited to, preparation of specifications and the construction or purchase of new production equipment; preparation of operator instructions; scheduling and control plans for production supplies; laboratory analysis support; data processing support; and clerical support.

1.4.4 Operator Quality Education

Expenditures incurred in the development and conduct of formal operator training programs for the expressed purpose of preventing errors—programs that emphasize the value of quality and the role that each operator plays in its achievement. This includes operator training programs in subjects such as statistical quality control, process control, quality circles, and problem-solving techniques. (This item is not intended to include any portion of basic apprentice or skill training necessary to be qualified for an individual assignment within a company.)

1.4.5 Operator SPC/Process Control

Expenditures incurred for education to implement the program.

1.5 Quality Administration

Expenditures incurred in the overall administration of the quality management function.

1.5.1 Administrative Salaries

Compensation costs for all quality function personnel such as managers and directors, supervisors, and clericals whose duties are 100% administrative.

1.5.2 Administrative Expenses

All other costs and expenses charged to or allocated to the quality management function not specifically covered elsewhere in this system (such as heat, light, telephone, and so on).

15.3 **Quality Program Planning**

The cost of quality (procedure) manual development and maintenance, inputs to proposals, quality record keeping, strategic planning, and budget control.

1.5.4 **Quality Performance Reporting**

Expenditures incurred in quality performance data collection, compilation, analysis, and issuance in report forms designed to promote the continued improvement of quality performance. Quality cost reporting would be included in this category.

1.5.5 **Quality Education**

Expenditures incurred in the initial (new employee indoctrination) and continued quality education of all company functions that can affect the quality of product or service as delivered to customers. Quality education programs emphasize the value of quality performance and the role that each function plays in its achievement.

1.5.6 **Quality Improvement**

Expenditures incurred in the development and conduct of companywide quality improvement programs designed to promote awareness of improvement opportunities and provide unique individual opportunities for participation and contributions.

1.5.7 **Quality System Audits**

Expenditures for audits performed to observe and evaluate the overall effectiveness of the quality management system and procedures. Often accomplished by a team of management personnel. Auditing of products is a monitoring cost. (See 2.2.1.)

1.6 Other Prevention Investments

Represents all other expenses of the quality system, not previously covered, specifically designed to prevent poor quality of product or service.

2.0 MONITORING COSTS

Costs associated with measuring, evaluating, or auditing products or services to assure conformance to quality standards and performance requirements.

2.1 Purchasing Monitoring Costs

Purchasing monitoring costs generally can be considered to be the costs incurred for the inspection and/or test of purchased supplies or services to determine acceptability for use. These activities can be performed as part of a receiving inspection function or as a source inspection at the supplier's facility.

2.1.1 Receiving or Incoming Inspections and Tests

Total costs for all normal or routine inspection and/or test of purchased materials, products, and services. These costs represent the baseline costs of purchased goods monitoring as a continuing part of a normal receiving inspection function.

2.1.2 Measurement Equipment

The cost of acquisition (depreciation or expense costs), calibration, and maintenance of measurement equipment, instruments, and gauges used for measurement or monitoring of purchased supplies.

2.1.3 Qualification of Supplier Product

The cost of additional inspections or tests (including environmental tests) periodically required to qualify the use of production quantities of purchased goods. These costs are usually one-time costs but they may be repeated during multi-year production situations. The following are typical applications:

a. First article inspection (detailed inspection and worst-case tests) on a sample of the first production buy of new components, materials, or services.

b. First article inspection for second and third sources of previously qualified end-product key components.

c. First article inspection of the initial supply of customer-furnished parts or materials.

d. First article inspection of the initial purchased quantity of goods for resale.

2.1.4 Source Inspection and Control Programs

All company-incurred costs (including travel) for the conduct of any of the activities described in 2.1.1 and 2.1.3 at the supplier's plant or at an independent test laboratory. This item will normally include all monitoring costs associated with direct shipments from supplier to the customer, sales office, or installation site.

2.2 Operations (Manufacturing or Service) Monitoring Costs

Operations monitoring costs generally can be considered to be the costs incurred for the inspections, tests, or audits required to determine and assure the acceptability of product or service to continue into each discrete step in the operations plan from start of production to delivery. In each case where material losses are an integral part of the monitoring operation, such as machine set-up pieces or destructive testing, the cost of the losses is to be included.

2.2.1 Planned Operations, Inspections, Tests, Audits

The cost of all planned inspections, tests, and audits conducted on product or service at selected points or work areas throughout the overall operations process including the point of final product or service acceptance. Also includes the total cost of any destructive test samples required. This is the baseline operations monitoring cost. It does not include the cost of troubleshooting, rework, repair, or sorting rejected lots, all of which are defined as waste costs.

2.2.1.1 Checking Labor

Work performed by individuals other than inspectors as in-process evaluation. Typically part of a production operator's job.

2.2.1.2 Product or Service Quality Audits

Personnel expense as a result of performing quality audits on in-process or finished products or services.

2.2.1.3 Inspection and Test Materials

Materials consumed or destroyed in control of quality, e.g., by tear-down inspections, over-voltage stressing, drop testing, or life testing.

2.2.2 Set-Up Inspections and Tests

The cost of all set-up or first piece inspections and tests utilized to assure that each combination of machine and tool is properly adjusted to produce acceptable products before the start of each production lot, or that service processing equipment (including acceptance and test devices) is acceptable for the start of a new day, shift, or other time period.

2.2.3 Special Tests (Manufacturing)

The cost of all non-routine inspections and tests conducted on manufactured product as a part of the monitoring plan. These costs normally include annual or semi-annual sampling of sensitive product for more detailed and extensive evaluations to assure continued conformance to critical environmental requirements.

2.2.4 Process Control Measurements

The cost of all planned measurements conducted on in-line product or service processing equipment and/or materials (e.g., oven temperature or material density) to assure conformance to pre-established standards. Includes adjustments made to maintain continued acceptable results.

2.2.5 Laboratory Support

The total cost of any laboratory tests required in support of product or service monitoring plans.

2.2.6 Measurement (Inspection and Test) Equipment

Since any measurement or process control equipment required is an integral part of monitoring operations, its acquisition (depreciation or expense), calibration, and maintenance costs are all included. Control of this equipment assures the integrity of results, without which the effectiveness of the appraisal program would be in jeopardy.

2.2.6.1 Depreciation Allowances

Total depreciation allowances for all capitalized monitoring equipment.

2.2.6.2 Measurement Equipment Expenses

The procurement or build cost of all monitoring equipment and gauges that are not capitalized.

2.2.6.3 Maintenance and Calibration Labor

The cost of all inspections, calibration, maintenance, and control of monitoring equipment, instruments, and gauges used for the evaluation of support processes, products, or services for conformance to requirements.

2.2.7 Outside Endorsements and Certifications

The total cost of required outside endorsements or certifications such as Underwriter's Laboratory, ASTM, or an agency of the U.S. government. Includes the cost of sample preparation, submittal, and any liaison necessary to its final achievement. Includes cost of liaison with customers.

2.3 External Monitoring Costs

External monitoring costs will be incurred any time there is need for field set-up or installation and checkout prior to official acceptance by the customer. These costs are also incurred when there is need for field trials of new products or services.

Field Performance Evaluation

The total cost of all monitoring efforts (inspections, tests, audits, and monitoring support activities) planned and conducted at the site for installation and/or delivery of large, complex products or the conduct of merchandised services (e.g., repairs or leasing set-ups).

2.3.1 Special Product Evaluations

Includes life testing and environmental and reliability tests performed on production units.

2.3.2 Evaluation of Field Stock and Spare Parts

Includes cost of evaluation testing or inspection of field stock, resulting from engineering changes, storage time (excessive shelf life), or other suspected problems.

2.4 Review of Test and Inspection Data

Costs incurred for regularly reviewing inspection and test data prior to release of the product for shipment, such as determining whether product requirements have been met.

2.5 Miscellaneous Quality Evaluations

The cost of all support area quality evaluations (audits) to assure continued ability to supply acceptable support to the production process. Examples of areas included are stores, packaging, and shipping.

3.0 WASTE COSTS

Costs resulting from products or services not conforming to requirements or customer/user needs. Waste costs occur prior to delivery or shipment of the product, or the furnishing of a service, to the customer.

3.1 Product/Service Design Waste Costs (Internal)

Design waste costs can generally be considered to be the unplanned costs that are incurred because of inherent design inadequacies in released documentation for production operations. They do not include billable costs associated with customer-directed changes (product improvements) or major redesign efforts (product upgrading) that are part of a company-sponsored marketing plan.

3.1.1 Design Corrective Action

After initial release of design for production, the total cost of all problem investigation and redesign efforts (including requalification as necessary) required to completely resolve product or service problems inherent in the design. (Some sources consider this an investment expense.)

3.1.2 Rework Due to Design Changes

The cost of all rework (materials, labor, and applicable burden) specifically required as part of design problem resolutions and implementation plan for required design changes.

3.1.3 Scrap Due to Design Changes

The cost of all scrap (materials, labor, and applicable burden) required as part of design problem resolutions and implementation plan for design changes.

3.1.4 Production Liaison Costs

The cost of unplanned production support efforts required because of inadequate or incomplete design description and documentation by the design organization.

3.2 Purchasing Waste Costs

Costs incurred due to purchased item rejects.

3.2.1 Purchased Material Reject Disposition Costs

The cost to dispose of, or sort, incoming inspection rejects. Includes the cost of reject documentation, review and evaluation, disposition orders, handling, and transportation (except as charged to the supplier).

3.2.2 Purchased Material Replacement Costs

The added cost of replacement for all items rejected and returned to the supplier. Includes additional transportation and expedition costs (when not paid for by the supplier).

3.2.3 Supplier Corrective Action

The cost of company-sponsored failure analyses and investigations into the cause of supplier rejects to determine necessary corrective actions. Includes the cost of visits to supplier plants for this purpose and the cost to provide necessary added inspection protection while the problem is being resolved. (Some sources consider this an investment expense.)

3.2.4 Rework of Supplier Rejects

The total cost of necessary supplier item repairs incurred by the company and not billable to the supplier—usually due to production time pressure.

3.2.5 Uncontrolled Material Losses

The cost of material or parts shortages due to damage, theft, or other (unknown) reasons. A measure of these costs may be obtained from reviews of inventory adjustments.

3.3 Operations (Product or Service) Waste Costs

Operations waste costs almost always represent a significant portion of overall quality costs and can generally be viewed as the costs associated with defective product or service discovered during the operations process. They may be categorized into three distinct areas: material review and corrective action, rework/repair costs, and scrap costs.

3.3.1 Material Review and Corrective Action Costs

Costs incurred in the review and disposition of nonconforming product or service and the corrective actions necessary to prevent recurrence.

3.3.1.1 Disposition Costs

All costs incurred in the review and disposition of nonconforming product or service, in the analysis of quality data to determine significant areas for corrective action, and in the investigation of these areas to determine the root causes of the defective product or service.

3.3.1.2 Troubleshooting or Failure Analysis Costs (Operations)

The cost of failure analysis (physical or chemical) conducted by or obtained from outside laboratories in support of defect cause identification. (Some sources consider this an investment expense.)

3.3.1.3 Investigation Support Costs

The additional cost of special runs of product or controlled lots of material (designed experiments) conducted specifically to obtain information useful to the determination of the root cause of a particular problem. (Some sources consider this an investment expense.)

3.3.1.4 Operations Corrective Action

The actual cost of corrective actions taken to remove or eliminate the root causes of nonconformances identified for correction. This item can include such activities as rewriting operator instructions, redevelopment of specific processes or flow procedures, redesign or modification of equipment or tooling, and the development and implementation of specific training needs. Does not include design (3.1.1) or supplier (3.2.3) corrective action costs. (Some sources consider this an investment expense.)

3.3.2 Operations Rework and Repair Costs

The total cost (labor, material, and overhead) of reworking or repairing defective product or service discovered within the operations process.

3.3.2.1 Rework

The total cost (material, labor, and burden) of all work done to bring nonconforming product or service up to an acceptable (conforming) condition, as authorized by specific work order, blueprint, personal assignment, or a planned part of the standard operating process. Does not include rework due to design change (3.1.2).

3.3.2.2 Repair

The total cost (material, labor, and burden) of all work done to bring non-conforming product up to an acceptable or equivalent, but still non-conforming, condition; normally accomplished by subjecting the product to an approved process that will reduce but not completely eliminate the nonconformance.

3.3.3 Re-inspection/Re-test Costs

That portion of inspection, test, and audit labor that is incurred because of rejects (includes documentation of rejects, re-inspection or test after re-work/repair, and sorting of defective lots).

3.3.4 Extra Operations

The total cost of extra operations such as touch-up or trimming, added because the basic operation is not able to achieve conformance to requirements. These costs are often hidden in the accepted (standard) cost of operations.

3.3.5 Operations Scrap Costs

The total cost (material, labor, and overhead) of defective product or service that is wasted or disposed of because it cannot be reworked to conform to requirements. The unavoidable losses of material (such as the turnings from machining work or the residue in a food mixing pot) are generally known as inherent process losses (check company cost accounting definitions) and are not to be included in the cost of quality. Also, in the definition of quality costs, the amount received from the sale of scrap and other salvage value is not to be deducted from gross scrap waste costs.

3.3.6 Downgraded End-Product or Service

Price differential between normal selling price and reduced selling price due to nonconforming or off-grade end-products or services because of quality reasons. Also includes any costs incurred to bring up to saleable condition.

3.3.7 Internal Waste Labor Losses

When labor is lost because of nonconforming work, there may be no concurrent material losses and it is not reflected on scrap or rework reports. Accounting for the cost of labor for such losses is the intent of this item. Typical losses occur because of equipment shutdowns and reset-up or line stoppages for quality reasons and may be efficiency losses or even allocated for by "labor allowances."

3.4 Other Waste Costs

3.5 Office Wastes

3.5.1 Meeting Waste

Meetings where no useful information was imparted, no one contributed information, and no progress was made on the work at hand. Examples: long rambling meetings without new information, meetings without agendas, meetings that allow discussions to wander from the topic at hand, etc.

3.5.2 Distractions

Activities not related to regular/ assigned work, but not including distractions that help create an insight or solution. Examples: browsing non-work internet sites such as social networking sites, unscheduled interruptions that prevented you from being productive, extra time on breaks and lunches, daydreaming, etc.

3.5.3 Messaging

Unproductive time spent on emails, phone calls, and texting. Examples: handling emails that you do not want or need, moving and filing emails, searching for old emails, dealing with emails that are rejected, and calling people, listening to messages, or texting messages where nothing useful is exchanged.

3.5.4 Information Loss from Storage and Retrieval Issues

Wasted time due to waiting for a system or software. Examples: poorly designed file areas such as a shared computer drive or server, searching for information or records, and requesting/ delivering replacement of lost information.

3.5.5 Engagement

Lacking personal motivation. Examples: lack of team support or social interactions, lack of interest, inability to see personal rewards for full work, slower work due to a perceived or real lack of compensation, inadequate challenge, or too-hard challenge.

3.5.6 Poor Results/Workflow

Decisions and/or work involving wasted time. Examples: lack of information, lack of communications, hardware or software technical issues, or delays by suppliers, missing information, poor decision-making processes, or misleading communications, priority conflicts, unknown next steps, lack of responsible follow up, interfering politics, improper assignment of duties and responsibilities.

3.5.7 Physical Arrangements

Physical arrangements that impede work, may include office space or desk space. May include wasting time in searching. Examples: communication interference, walk time to reach a location, wasted trips to the office printer or supply cabinet, reaching for a piece of equipment.

4.0 DOWNSTREAM CONSEQUENCE COSTS

Costs resulting from products or services not conforming to requirements or customer/user needs. Downstream consequence costs occur after delivery or shipment of the product, and during or after furnishing of a service, to the customer.

4.1 Complaint Investigations/Customer or User Service

The total cost of investigating, resolving, and responding to individual customer or user complaints or inquiries, including necessary field service.

4.2 Returned Goods

The total cost of evaluating and repairing or replacing goods not meeting acceptance by the customer or user due to quality problems. It does not include repairs accomplished as part of a maintenance or modification contract.

4.3 Retrofit Costs

Costs to modify or update products or field service facilities to a new design change level based on major redesign due to design deficiencies. Includes only that portion of retrofits that are due to quality problems.

4.3.1 Recall Costs

Includes costs of recall activity due to quality problems.

4.4 Warranty Claims

The total cost of claims paid to the customer or user, after acceptance, to cover expenses. Includes repair costs such as removing defective hardware from a system or cleaning costs due to a food or chemical service accident. In cases where a price reduction is negotiated in lieu of warranty, the value of this reduction should be counted.

4.5 Liability Costs

Company-paid costs due to liability claims, including the cost of product or service liability insurance.

4.6 Penalties

Cost of any penalties incurred because of less than full product or service performance achieved (as required by contracts with customers and government rules and regulations).

4.7 Customer/User Goodwill

Costs incurred, over and above normal selling costs, to customers or users who are not completely satisfied with the quality of delivered product or service. Includes costs incurred because customer quality expectations are greater than product or service received.

4.8 Lost Sales

Includes value of contribution margin lost due to sales reduction because of quality problems. (These costs should be clearly and directly related to specific instances of poor quality, not surmised or assumed to be vaguely related to prior issues.)

4.9 Other Downstream Consequence Costs

Consolidated References

CHAPTER 1

American Society for Quality. "What Is Social Responsibility?" https://asq.org/quality-resources/social-responsibility.

ANSI/ASQC ISO/DIS10014. *Guidelines for Managing the Economics of Quality—Draft*. International Standard, 1996.

ANSI/ASQC Q9004-1-1994. *Quality Management and Quality System Elements—Guidelines*. ASQC Quality, 1994.

Antonaras, Alexandros, Christina Memtsa, and Melanthia Iacovidou. "The Challenge of Measuring the Cost of Quality." Paper presented at the Fourth Annual Quality Congress Middle East, Dubai, UAE, March 22–24, 2010.

Brown, F. X., and R. W. Kane. "Quality Costs and Profit Performance." In *Annual Technical Conference Transactions,* collectively edited by ASQ. American Society for Quality Control, 1975.

Byrne, Diane M., and Nancy E. Ryan, eds. *Taguchi Methods and QFD*. ASI, 1988.

Campanella, Jack, ed. *Principles of Quality Costs: Principles, Implementation, and Use*. 3rd ed. ASQ Quality, 1999.

Campanella, Jack, ed. *Quality Costs: Ideas and Applications*. Vol. 2. ASQ Quality, 1989.

Chartered Institute of Management Accountants. "Management Accounting Tools for Today and Tomorrow." Survey, July 2009.

Dawes, Edgar W. "Quality Costs—New Concepts and Methods." In *Annual Quality Congress Transactions,* collectively edited by ASQ. American Society for Quality Control, 1987.

Department of Defense. *MIL-Q-9858A, Quality Program Requirements*. Department of Defense, 1963.

Ealey, Lance A. *Quality by Design*. ASI, 1988.

Eureka, William E., and Nancy E. Ryan, eds. *The Customer-Driven Company*. ASI, 1988.

Feigenbaum, Armand V. *Total Quality Control: Engineering and Management*. McGraw-Hill, 1961.

Freeman, Harold L. "How to Put Quality Costs to Use." In *Transactions of the Metropolitan Conference*, collectively edited by ASQ. American Society for Quality Control, 1960.

Grimm, Andrew F., ed. *Quality Costs: Ideas and Applications*. Vol. 1. ASQC Quality, 1987.

Harrington, H. J. *Poor-Quality Cost*. ASQC Quality, 1987.

International Organization for Standardization. *ISO 10014:2006, Quality Management—Guidelines for Realizing Financial and Economic Benefits*. International Organization for Standardization, 2006.

International Organization for Standardization. *ISO 9000:2005, Quality Management Systems—Fundamentals and Vocabulary*. International Organization for Standardization, 2005.

International Organization for Standardization. *ISO 9001:2008, Quality Management Systems—Requirements*. International Organization for Standardization, 2008.

International Organization for Standardization. *ISO 9004:2009, Quality Management Systems—Managing for Sustained Success of an Organization*. International Organization for Standardization, 2009.

"ISO 9001:2000." *Quality Management Forum*, Winter 2006, 7–9.

Juran, J. M., ed. *Quality Control Handbook*. 4th ed. New York: McGraw-Hill, 1988.

Juran, J. M., A. Blanton Godfrey, and Frank M. Gryna. "Section 8 Quality and Costs." In *Quality Control Handbook*, 5th ed., edited by J. M. Juran. McGraw-Hill, 1999.

Juran, J. M., and Frank M. Gryna. "Section 4, Quality Costs." In *Quality Control Handbook*, 4th ed., edited by J. M. Juran. New York: McGraw-Hill, 1988.

Kaner, Cem. "Quality Cost Analysis: Benefits and Risks." Paper presented at the Software Management Conference, Santa Clara, CA, January 1996.

Liebesman, Sandford. "Continual Improvement Using ISO 9001." *Quality Progress*, January 2003, 62–63.

Liebesman, Sandford. "Increase ISO 9001's Value." *Quality Progress*, August 2006, 84–85.

Liebesman, Sandford. "TL 9000: The Telecommunications Quality Management System." In *The ASQ ISO 9000:2000 Handbook*, edited by Charles A. Cianfrani, Joseph J. Tsiakals, and John E. West. ASQ Quality, 2002.

Liebesman, Sandford, Alka Jarvis, and Ashok V. Dandekar. *TL 9000, Release 3.0, and a Guide to Measuring Excellence in Telecommunications*. 2nd ed. ASQ Quality, 2002.

Liebesman, Sandford, and John Walz. "New Version of TL 9000 Released." *Standards Outlook, Quality Progress,* March 2007, 74–76.

Masser, W. J. "The Quality Manager and Quality Costs." *Industrial Quality Control* 14, no. 4 (1957): 5–8.

QS-9000. Chrysler Corporation, Ford Motor Company, and General Motors Corporation. 1998. *QS-9000 Quality System Requirements.* 3rd ed. Southfield, MI: Automotive Industry Action Group.

QuEST Forum. *TL 9000 Requirements Handbook, Release 5.0* QuEST Forum, 2009.

QuEST Forum. *TL 9000 Measurements Handbook, Release 4.5.* QuEST Forum, 2010.

Raßfeld, Christian, Falk Behmer, Marc Dürlich, and Roland Jochem. "Do Quality Costs Still Matter?" *Total Quality Management and Business Excellence* 26, nos. 9–10 (2015): 1071–82. https://doi.org/10.1080/14783363.2015.1068591.

Rodchua, Suhansa. "Quality Costs and Enterprise Size in the Manufacturing Industry: A Survey of American Society for Quality (ASQ) Members." PhD diss., Indiana State University, 2005. https://scholars.indianastate.edu/etds/1045.

Schiffauerova, Andrea, and Vince Thomson. "Managing Cost of Quality: Insight into Industry Practice." *TQM Magazine* 18, no. 5 (2006): 457–81.

Setijono, Djoko, and Jens J. Dahlgaard. *The Value of Quality Improvements.* Växjö University and Linköping University, 2008.

Sower, Victor E., and Ross Quarles. "Cost of Quality Usage and Its Relationship to Quality Systems Maturity." Working paper series, Center for Business and Economic Development at Sam Houston State University, November 2002.

Sower, Victor E., and Ross Quarles. *Cost of Quality: Why More Organizations Do Not Use It Effectively*. American Society for Quality, 2003.

Teli, S. N., V. S. Majali, U. M. Bhushi, and S. Gaikwaid. "Daily Management Process for Reduce Cost of Quality in Automobile Industry." *Journal of Mechanical Engineering and Technology* (*JMET*) 1, no. 1 (2013): 45–53.

"The National Conference for Quality." *Quality Progress* 15, no. 5 (May 1952): 14–17.

Wood, Douglas C., ed. *Principles of Quality Costs: Its Principles, Implementation, and Use*. 4th ed. ASQ Quality, 2013.

Wood, Douglas C. "Types of Waste in the Office and Their Relative Magnitudes." Master's thesis, University of Central Missouri, 2014.

Yang, Chih-Chuang. "Improving the Definition and Quantification of Quality Costs." *Total Quality Management and Business Excellence* 19, no. 3 (2008): 175–91.

CHAPTER 2

Campanella, Jack, and Frank J. Corcoran. "Principles of Quality Costs." In *ASQC Quality Congress Transactions—Detroit*, collectively edited by ASQ. American Society for Quality Control, 1982.

Dale, Barrie G., and James J. Plunkett. *Quality Costing*. 2nd ed. Chapman & Hall, 1995.

Gryna, Frank M. "Quality Costs." In Juran, *Quality Control Handbook.*

Schiffauerova, Andrea, and Vince Thomson. "Managing Cost of Quality: Insight into Industry Practice." *TQM Magazine* 18, no. 5 (2006): 542–50. https://doi.org/10.1108/09544780610685502.

CHAPTER 3

Al-Dujaili, M. A. A. "Study of the Relation Between Types of the Quality Costs and Its Impact on Productivity and Costs: A Verification in Manufacturing Industries." *Total Quality Management and Business Excellence* 24, nos. 3–4 (2013): 397–419.

American Society for Quality. *Technical Report 2 (TR2): Cost of Quality—Guidelines for Development, Implementation, and Monitoring to Improve Quality and Performance.* American Society for Quality, 2024.

Atkinson, J. H., G. Hohner, B. Mundt, R. B. Troxel, and W. Winchell. *Current Trends in Cost of Quality.* National Association of Accountants, 1991.

Ayach, L., A. Anouar, and M. Bouzziri. "Quality Cost Management in Moroccan Industry: QMS Maturity and Sales Impact." *International Journal of Quality and Reliability Management* 36, no. 8 (2019): 1447–65.

Campanella, Jack. *Principles of Quality Costs.* 2nd ed. ASQC Quality, 1990.

da Silva Neto, J. A., G. E. O. Giacaglia, and W. de Queiróz Lamas. "ABC Quality-Cost Model: A Case Study." *TQM Journal* 31, no. 6 (2019): 928–47.

Freeman, J. M. "The Case for Quality Costing Simulation." *TQM Journal* 20, no. 6 (2008): 556–67.

Juran, J. M., and Frank M. Gryna. *Quality Planning and Analysis.* 3rd ed. McGraw-Hill, 1993.

Ramu, G. "Standard Issues: ISO 10014 & ASQ TR 2—Linking CoQ to Business Results." *Quality Progress* 54, no. 8 (2021): 24–29.

Snee, R. D., and R. W. Hoerl. "Holistic Improvement: From Lean Six Sigma to Statistical Engineering and Big Data." *Quality Progress* 51, no. 8 (2018): 34–40.

Webster, D. W. "Achieving Value Through Activity-Based Costing." In Proceedings of the ASQC Forty-Ninth Annual Quality Congress, Cincinnati, OH, 1995.

CHAPTER 4

American Society for Quality. "Software Quality Engineer (CSQE) Certification." Accessed January 18, 2026. https://www.asq.org/cert/software-quality-engineer.

American Society for Quality Control, Quality Costs Committee. *Guide for Reducing Quality Costs*. 2nd ed. ASQC Quality, 1987.

Briscoe, Nat R., and Frank M. Gryna. *Assessing the Cost of Poor Quality in a Small Business*. Report No. 902, University of Tampa College of Business, May 1996.

Dale, Barrie G., and James J. Plunkett. *Quality Costing*. 2nd ed. Chapman & Hall, 1995.

Dion, R. "Process Improvement and the Corporate Balance Sheet." *IEEE Software* 10, no. 4 (1993): 28–35. https://doi.org/10.1109/52.219618.

Futrell, Robert T., Donald F. Shafer, and Linda I. Shafer. *Quality Software Project Management*. Prentice Hall PTR, 2002.

Haley, T. J. "Software Process Improvement at Raytheon." *IEEE Software* 13, no. 6 (1996): 33–41. https://doi.org/10.1109/52.542292.

Kristensen, Kai, and Jacob Eskildsen. "The Relationship Between SERVQUAL, National Customer Satisfaction Indices, and Consumer Sentiment." *Quality Management Journal* 19, no. 2 (2012): 47–61. https://doi.org/10.1080/10686967.2012.11918346.

Löfgren, Martin, and Lars Witell. "Two Decades of Using Kano's Theory of Attractive Quality: A Literature Review." *Quality Management Journal* 15, no. 1 (2008): 59–75. https://doi.org/10.1080/10686967.2008.11918056.

Paulk, Mark C., Bill Curtis, Mary Beth Chrissis, and Charles V. Weber. *Capability Maturity Model for Software, Version 1.1*. Technical Report CMU/SEI-93-TR-24. Software Engineering Institute, Carnegie Mellon University, February 1993.

Phillips, Jack. "Valuing Quality Initiatives: New Perspectives for Measuring Success with Six Types of Data." *Quality Management Forum* 34, no. 8 (2008): 21–24.

Schiffauerova, Alena, and Vaughan Thomson. "Managing Cost of Quality: Insight into Industry Practice." *TQM Magazine* 18, no. 5 (2006): 542–50. https://doi.org/10.1108/09544780610685502.

Stamatis, D. H. *Failure Mode and Effect Analysis: FMEA from Theory to Execution*. 2nd rev. and expanded ed. ASQ Quality, 2003.

Teplická, Katarína, and Soňa Hurná. "New Approach of Costs of Quality According Their Trend of During Long Period in Industrial Enterprises in SMEs." *Management Systems in Production Engineering* 29, no. 1 (2021): 20–26. https://doi.org/10.2478/mspe-2021-0003.

CHAPTER 5

Al-Dujaili, M. A. A. "Study of the Relation Between Types of the Quality Costs and Its Impact on Productivity and Costs: A Verification in Manufacturing Industries." *Total Quality Management and Business Excellence* 24, nos. 3–4 (2013): 397–419.

American Society for Quality. *Technical Report 2 (TR2): Cost of Quality—Guidelines for Development, Implementation, and Monitoring to Improve Quality and Performance*. ASQ, 2024.

ASQ Quality Costs Committee. *Guide for Reducing Quality Costs*. 2nd ed. ASQC Quality, 1987.

ASQE/IOE. ASQE/IOE 2025 survey. 2025.

Atkinson, John Hawley, Jr., Gregory Hohner, Barry Mundt, Richard B. Troxel, and William Winchell. *Current Trends in Cost of Quality: Linking the Cost of Quality and Continuous Improvement*. National Association of Accountants, 1991.

Ayach, L., A. Anouar, and M. Bouzziri. "Quality Cost Analysis in Cement Industry: Case Study." *International Journal of Quality and Reliability Management* 35, no. 9 (2018): 2018–35.

da Silva Neto, J. A., G. E. O. Giacaglia, and W. de Queiróz Lamas. "ABC Quality-Cost Model: A Case Study." *TQM Journal* 31, no. 6 (2019): 928–47.

Holota, T., M. Sarnovsky, and K. Balog. "Management of Quality Costs: Analysis and Methods." *Quality Innovation Prosperity* 19, no. 1 (2015): 22–36.

Juran, J. M., and Frank M. Gryna. *Quality Planning and Analysis: From Product Development Through Use*. 3rd ed. McGraw-Hill, 1993.

Juran, J. M., and Frank M. Gryna. "Section 22, Quality Improvement." In Juran, *Quality Control Handbook*.

Mitreva, E., H. Gjorshevski, and N. Taskov. "The Need for Analysis of Costs of Quality." *International Journal of Information, Business and Management* 10, no. 2 (2018): 98–112.

Newnan, Donald G. *Engineering Economic Analysis*. 4th ed. Engineering Press, 1991.

Phillips, J. "Valuing Quality Initiatives: Six Types of Data." Quality Progress 41, no. 2 (2008): 50–55.

Quality Cost Committee. *Principles of Quality Costs*. 2nd ed. Edited by Jack Campanella. ASQC Quality, 1990.

Snee, R. D., and R. W. Hoerl. "Holistic Improvement: From Lean Six Sigma to Statistical Engineering and Big Data." *Quality Progress* 51, no. 8 (2018): 34–40.

Winchell, William. *Continuous Quality Improvement: A Manufacturing Professionals Guide*. Society of Manufacturing Engineers, 1991.

Winchell, William. "Driving Buyer Satisfaction by Quality Cost." In *Fiftieth Annual Quality Congress Transactions*, collectively edited by ASQ. American Society for Quality Control, 1996.

CHAPTER 6

ASQ Quality Costs Committee. *Guide for Managing Supplier Quality Costs.* 2nd ed. ASQC Quality, 1987.

Byrne, Diane M., and Nancy E. Ryan, eds. *Taguchi Methods and QFD*. ASI, 1988.

Ealey, Lance A. *Quality by Design*. ASI, 1988.

Eureka, William E., and Nancy E. Ryan, eds. *The Customer-Driven Company.* ASI, 1988.

Pattanayak, A. K., A. Prakash, and R. P. Mohanty. "Managing Cost of Quality in Supply Chain." *International Journal of Quality and Reliability Management* 36, no. 7 (2019): 1223–43.

CHAPTER 7

Brodman, J., and D. Johnson. "Return on Investment from Software Process Improvement as Measured by U.S. Industry." *CrossTalk: Journal of Defense Software Engineering* 9, no. 4 (1996): 23–28.

Dion, R. "Cost of Quality as a Measure of Process Improvement." Proceedings of the SEI Software Engineering Symposium, Pittsburgh, PA September 17, 1992.

Dion, R. "Process Improvement and the Corporate Balance Sheet." *IEEE Software* 10, no. 4 (1993): 28–35.

Dion, R. "Quantifying the Benefit of Software Process Improvement." Proceedings of the SEI / ALAA Software Process Improvement Workshop, Chantilly, VA, November 8, 1990.

Futrell, R., D. Shafer, and L. Shafer. *Quality Software Project Management.* Prentice Hall, 2002.

Gack, G. "Measuring Software Defect Containment." *Quality Progress* 44, no. 8 (2011): 34–41.

Gack, Gary. "Powerful Metrics: Software Cost of Quality + Defect Containment." *Software Quality Professional* 13, no. 2 (2011): 5–8.

Gibson, Diane, Dennis Goldenson, and Keith Kost. *Performance Results of CMMI-Based Process Improvement (CMU/SE1-2006-TR-004).* Software Engineering Institute, Carnegie Mellon University, 2006. http: / / www .sei.cmu.edu / library / abstracts / reports / 06tr004.cfm.

Haley, T., B. Ireland, E. Wojtaszek, D. Nash, and R. Dion. "Raytheon Electronic Systems Experience in Software Process Improvement."

Technical Report CMU/SEI-95-TR-017, November 1995. Published by the Software Engineering Institute (SEI) at Carnegie Mellon University.

Haley, T.J. "Software Process Improvement at Raytheon." *IEEE Software* 13, no. 6 (1996): 33–41.

International Organization for Standardization. *ISO/IEC 25010:2011: Systems and software engineering—Systems and software Quality Requirements and Evaluation (SQuaRE)—System and Software Quality Models*. International Organization for Standardization, 2011.

International Organization for Standardization, International Electrotechnical Commission, and Institute of Electrical and Electronics Engineers. *ISO/IEC/IEEE 90003:2018: Software Engineering—Guidelines for the Application of ISO 9001:2015 to Computer Software*. ISO/IEC/IEEE, 2018.

Jones, C. "The Economics of Software Quality." *Quality Progress* 44, no. 10 (2011): 46–53.

Jones, Capers. "Software Quality and Software Costs." *Software Quality Professional* 13, no. 3 (2011): 28–29.

Karg, Lars M., Michael Grottke, and Ame Beckhaus. "A Systematic Literature Review of Software Quality Cost Research." *Journal of Systems and Software* 84, no. 3 (2011): 415–27.

Knox, S. T. "Modeling the Cost of Software Quality." *Digital Technical Journal* 5, no. 4 (1993): 9–16.

Krasner, H. "Accumulating the Body of Evidence for the Payoff of SPI." University of Texas, 1997. https://www.utexas.edu/coe/cqi/archive.

Kristensen, K., and J. Eskildsen. "The Relationship Between SERVQUAL, National Customer Satisfaction Indices, and Consumer Sentiment." *ASQ Quality Management Journal* 19, no. 2 (2012): 50–55.

Liu, A. Q. "Motorola Software Group's China Center: Value Added by CMMI." *Software Technology News* 10, no. 1 (2007): 19–23. http://www.journal.thedacs.com/issue/41/79.

Lofgren, M., and L. Witell. "Two Decades of Using Kano's Theory of Attractive Quality: A Literature Review." *ASQ Quality Management Journal* 15, no. 1 (2008): 61–64.

Mahanti, R. "Measuring Cost of Software Quality." *Quality Progress* 44, no. 5 (2011): 23–29.

Mahanti, Rupa. "Software Six Sigma and Cultural Change: The Key Ingredients." *Software Quality Professional* 13, no. 2 (March 2011): 43–45.

Mulcahy, R., and M. Kirwin. *PMP Exam Prep: Accelerated Learning to Pass the Project Management Professional (PMP) Exam*. 11th ed. RMC, 2023.

Phillips, Jack. "Valuing Quality Initiatives: New Perspectives for Measuring Success with Six Types of Data." *Quality Management Forum* 34, no. 8 (Fall 2008): 21–24.

Pressman, Roger S. *Software Engineering: A Practitioner's Approach*. 6th ed. McGraw-Hill, 2005.

Price Waterhouse Management Consultants. *Software Quality Standards: The Costs and Benefits. A Review for the Department of Trade and Industry*. Price Waterhouse Management Consultants, 1988.

Slaughter, Sandra A., Donald E. Harter, and Mayuram S. Krishnan. "Evaluating the Cost of Software Quality." *Communications of the ACM* 41, no. 8 (1998): 67–73.

Sutherland, J., and J. J. Sutherland. *Scrum: The Art of Doing Twice the Work in Half the Time*. Crown Business, 2014.

van Solingen, Rini. "Measuring the ROI of Software Process Improvement." *IEEE Software* 21, no. 3 (2004): 32–38.

Weller, E.F. "Lessons from Three Years of Inspection Data." *IEEE Software* 10, no. 5 (1993): 38–45.

Westfall, L. V. *The ASQ Certified Software Quality Engineer Handbook*. 3rd ed. ASQ Quality, 2024.

CHAPTER 8

Dion, R. 1990. "Quantifying the Benefit of Software Process Improvement." Paper presented at the SEI/AIAA Software Process Improvement Workshop, Chantilly, VA, November 8.

Galati, Armand J. "Cost of Quality in a University Environment." *Quality Progress* 26, no. 10 (1993): 101–6.

Haley, Thomas J., Raylene Swanson, Neil Rogers, and Edward R. Baker. "Software Process Improvement at Raytheon." *IEEE Software* 12, no. 6 (1995): 33–41. https://doi.org/10.1109/52.469758.

Harrington, H. James. *The Poor Quality Cost*. Marcel Dekker, 1987.

Juran, J. M., and Frank M. Gryna. *Quality Planning and Analysis: From Product Development Through Use*. 3rd ed. McGraw-Hill, 1993.

Krasner, Herb. "Accumulating the Body of Evidence for the Payoff of Software Process Improvement." In *Software Process Improvement*, edited by Robin B. Hunter and Richard H. Thayer. IEEE Computer Society, 2001.

Sambamurthy, V., and Robert W. Zmud. "Banc One Corporation: Measuring the Cost of Quality." IT Management Case Study Series, 1994.

Wood, Douglas C. "Applying Lean Principles to Clinic Operations, Part 1." *Nephrology News and Issues* 25, no. 12 (2011): 24–27.

Wood, Douglas C. "Applying Lean Principles to Clinic Operations, Part 2." *Nephrology News and Issues* 25, no. 13 (2011): 20–22.

Wood, Douglas C. "Project Oriented Cost of Quality." *Quality Management Forum*, Winter 2008, 1, 4–5.

Wood, Douglas C. "Types of Waste in the Office and Their Relative Magnitudes." Master's thesis, University of Central Missouri, 2014.

CHAPTER 9

American Society for Quality (ASQ). *TR 2–2024: Practical Guidance for COQ Programs*. ASQ, 2024.

Crotty, Brendan. *Advanced QMS For Dummies*. 3rd ETQ special ed. John Wiley & Sons, 2018.

International Automotive Task Force. *IATF 16949: Quality Management System Requirements for Automotive Production and Relevant Service Parts Organizations*. IATF, n.d.

International Organization for Standardization. *ISO 10014: Quality Management—Guidelines for Realizing Financial and Economic Benefits*. ISO, n.d.

International Organization for Standardization. *ISO 9001:2015: Quality Management Systems—Requirements*. 5th ed. ISO, 2015.

CHAPTER 10

Al-Dujaili, M. A. A. "Study of the Relation Between Types of the Quality Costs and Its Impact on Productivity and Costs: A Verification in Manufacturing Industries." *Total Quality Management and Business Excellence* 24, nos. 3–4 (2013): 397–419.

Carr, Lawrence P. "Applying Cost of Quality to a Service Business." *MIT Sloan Management Review* (July 1992): 72–77.

Damodaran, Aswath. *Narrative and Numbers: The Value of Stories in Business*. Columbia University Press, 2017.

da Silva Neto, J. A., Giorgio E. O. Giacaglia, and W. de Queiróz Lamas. "ABC Quality-Cost Model: A Case Study." *TQM Journal* 31, no. 6 (2019): 928–47.

Dion, R. "Process Improvement and the Corporate Balance Sheet." *IEEE Software* 10, no. 4 (1993): 28–35.

Engelen, E., I. Ertürk, J. Froud, and K. Williams. "Reconceptualizing Financial Innovation: Frame, Conjuncture and Bricolage." *Economy and Society* 39, no. 1 (2010): 33–63. https://doi.org/10.1080/03085140903424568.

Ertürk, I. *Post-Crisis Central Bank Unconventional Policies and Financialised Transmission Channels*. Foundation for European Progressive Studies, 2016.

Ertürk, I., J. Froud, S. Johal, A. Leaver, and K. Williams. "Against Agency: A Positional Critique." *Economy and Society* 36, no. 1 (2007): 51–77. https://doi.org/10.1080/03085140601089903.

Fama, Eugene F. "Agency Problems and the Theory of the Firm." *Journal of Political Economy* 88, no. 2 (1980): 288–307. http://www.jstor.org/stable/1837292.

Foroohar, Rana. "Too Many Businesses Want a Piece of the Financial Action." *Financial Times*, May 15, 2016. https://www.ft.com/content/ed421ea4-1925-11e6-b197-a4af20d5575e.

Freeman, J. M. "The Case for Quality Costing Simulation." *TQM Journal* 20, no. 6 (2008): 556–67.

Friedman, Milton. *Capitalism and Freedom*. University of Chicago Press, 1962.

Groves, Martha. "IBM Stops Shipping Pentium PCs." *Los Angeles Times*, December 13, 1994.

Intel Corporation. "Intel 1994 Revenue, Earnings per Share Set Records." Press release, January 17, 1995.

Jensen, Michael C. "Organization Theory and Methodology." *Accounting Review* 58, no. 2 (1983): 319–39. http://www.jstor.org/stable/246838.

Kaplan, Robert S., and David P. Norton. "The Balanced Scorecard: Measures That Drive Performance." *Harvard Business Review* 70, no. 1 (1992): 71–79.

Kaplan, Robert S., and David P. Norton. *The Balanced Scorecard: Translating Strategy into Action*. Harvard Business Press, 1996.

Kaplan, Robert S., and David P. Norton. *The Strategy-Focused Organization: How Balanced Scorecard Companies Thrive in the New Business Environment*. Harvard Business Press, 2001.

Kearns, K. "Paul O'Neill: Cross-Sector Leadership Principles." Case study PDF, University of Pittsburgh, Johnson Institute, n.d.

National Institute of Standards and Technology. "1989 Malcolm Baldrige National Quality Award: Xerox Corporation Profile." PDF, 2001.

Nørreklit, Hanne. "The Balance on the Balanced Scorecard—A Critical Analysis of Some of Its Assumptions." *Management Accounting Research* 11, no. 1 (2000): 65–88.

Quanterix Corporation. Press releases, Form 8-K exhibits, and Q4 FY22 earnings materials, 2022–23. http://ir.quanterix.com.

Wigglesworth, Robin, Leo Lewis, and Dan McCrum. "Central Banks: Negative Thinking." *Financial Times*, February 17, 2016. https://www.ft.com/content/7333e92a-d4a2-11e5-829b-8564e7528e54.

Wolf, Martin. "Why Finance Is Too Much of a Good Thing." *Financial Times*, May 26, 2015. https://www.ft.com/content/64c2f03a-03a0-11e5-a70f-00144feabdc0.

Zrymiak, D. "Cost of Social Responsibility: Lessons from Automotive." *Quality Progress* 49, no. 5 (2016): 44–49.

APPENDIX A

Al-Dujaili, M. A. A. "Study of the Relation Between Types of the Quality Costs and Its Impact on Productivity and Costs: A Verification in Manufacturing Industries." *Total Quality Management and Business Excellence* 24, nos. 3–4 (2013): 397–419.

Asada, R. Y., Y. Kumar, and W. Al-Hubaishi. "A Review of Models for Costing Quality." *TQM Journal* 33, no. 7 (2021): 1425–48.

Ayach, L., Anouar, A., and Bouzziri, M. "Quality Cost Management in Moroccan Industry: QMS Maturity and Sales Impact." *International Journal of Quality and Reliability Management* 36, no. 8 (2019): 1447–65.

Blank, L. T., and A. J. Tarquin. *Engineering Economy*. 8th ed. McGraw-Hill Education, 2018.

Brealey, R. A., S. C. Myers, and F. Allen. *Principles of Corporate Finance*. 13th ed. McGraw-Hill Education, 2020.

Brigham, E. F., and M. C. Ehrhardt. *Financial Management: Theory and Practice*. 17th ed. Cengage Learning, 2022.

Damodaran, A. *Investment Valuation: Tools and Techniques for Determining the Value of Any Asset*. 3rd ed. Wiley, 2012.

da Silva Neto, J. A., Giacaglia, G. E. O., and de Queiróz Lamas, W. "ABC Quality-Cost Model: A Case Study." *TQM Journal* 31, no. 6 (2019): 928–47.

Kaplan, R. S., and D. P. Norton. *The Balanced Scorecard: Translating Strategy into Action*. Harvard Business School Press, 1996.

Kaplan, R. S., and D. P. Norton. *Strategy Maps: Converting Intangible Assets into Tangible Outcomes*. Harvard Business School Press, 2004.

Peterson, P. P., and F. J. Fabozzi. *Capital Budgeting: Theory and Practice*. Wiley, 2002.

Ramu, G. "Standard Issues: ISO 10014 and ASQ TR 2—Linking CoQ to Business Results." *Quality Progress* 54, no. 9 (2021): 56–59.

Reuer, J. J., and T. W. Tong. Real Options in International Joint Ventures. *Journal of International Business Studies* 41, no. 5 (2010): 807–26.

www.ingramcontent.com/pod-product-compliance
Lightning Source LLC
LaVergne TN
LVHW060028210926
848481LV00013B/3815
* 9 7 8 1 6 3 6 9 4 2 1 1 7 *